Studies in Computational Intelligence

Volume 726

Series editor

Janusz Kacprzyk, Polish Academy of Sciences, Warsaw, Poland
e-mail: kacprzyk@ibspan.waw.pl

About this Series

The series "Studies in Computational Intelligence" (SCI) publishes new developments and advances in the various areas of computational intelligence—quickly and with a high quality. The intent is to cover the theory, applications, and design methods of computational intelligence, as embedded in the fields of engineering, computer science, physics and life sciences, as well as the methodologies behind them. The series contains monographs, lecture notes and edited volumes in computational intelligence spanning the areas of neural networks, connectionist systems, genetic algorithms, evolutionary computation, artificial intelligence, cellular automata, self-organizing systems, soft computing, fuzzy systems, and hybrid intelligent systems. Of particular value to both the contributors and the readership are the short publication timeframe and the worldwide distribution, which enable both wide and rapid dissemination of research output.

More information about this series at http://www.springer.com/series/7092

Roger Lee

Editor

Computational Science/Intelligence and Applied Informatics

 Springer

Editor
Roger Lee
Software Engineering and Information
  Technology Institute
Central Michigan University
Mt. Pleasant, MI
USA

ISSN 1860-949X   ISSN 1860-9503   (electronic)
Studies in Computational Intelligence
ISBN 978-3-319-87596-5   ISBN 978-3-319-63618-4   (eBook)
DOI 10.1007/978-3-319-63618-4

Printed on acid-free paper

This Springer imprint is published by Springer Nature
The registered company is Springer International Publishing AG
The registered company address is: Gewerbestrasse 11, 6330 Cham, Switzerland

Foreword

The purpose of the 4th ACIS International Conference on Computational Science/Intelligence and Applied Informatics (CSII 2017) which was held on July 9–13, 2017 in Hamamatsu, Japan was to gather researchers, scientists, engineers, industry practitioners, and students to discuss, encourage and exchange new ideas, research results, and experiences on all aspects of Computational Science/Intelligence and Applied Informatics and to discuss the practical challenges encountered along the way and the solutions adopted to solve them. The conference organizers have selected the best 16 papers from those papers accepted for presentation at the conference in order to publish them in this volume. The papers were chosen based on review scores submitted by members of the program committee and underwent further rigorous rounds of review.

In Chapter "Proposed Framework Application for a Quality Mobile Application Measurement and Evaluation," Mechelle Grace Zaragonza and Haeng-Kon Kim describe an object-oriented (OO) approach to a software measurement framework aimed at evaluating software products, software process, and resources. This approach includes the dynamic characteristics in software measurement, considering the behavior aspects of software metrics.

In Chapter "Proposal and Development of Artificial Personality(AP) Application using the 'Requesting' Mechanism," Yosuke Kanai and Takayuki Fujimoto developed the "Artificial Personality (AP)" that is a designed form of a personality without autonomous learning ability, a software to express more "humanlike qualities" than AI by efforts for manual steps.

In Chapter "Load Experiment of the vDACS Scheme in Case of the 300 Simultaneous Connection," Kazuya Odagiri, Shogo Shimizu, and Naohiro Ishii perform a load experiment of the cloud type virtual PBNM named the vDACS Scheme, which can be used by plural organizations, for applications to the small and medium size scale organization.

In Chapter "Hearing-Dog Robot to wake People up Using its Bumping Action," Yukihiro Yoshida, Daiki Sekiya, Tsuyoshi Nakamura, Masayoshi Kanoh, and Koji Yamada propose a robot inspired by the behavior of hearing dogs. They conducted an experiment to evaluate the usefulness of the robot to wake up sleeping people.

In Chapter "Implementation of Document Production Support System with Obsession Mechanism," Ziran Fan and Takayuki Fujimoto consider the work efficiency in people's work style today, and focus on the task of document production. They focus on the most essential point in producing document, which is "to complete within a deadline."

In Chapter "Detecting Outliners in Terms of Errors in Embedded Software Development Projects Using Imbalance Data Classification," Kazunori Iwata, Toyoshiro Nakashima, Yoshiyuki Anan, and Naohiro Ishii examine the effect of undersampling on the detection of outliers in terms of the number of errors in embedded software development projects. Their study aims at estimating the number of errors and the amount of effort in projects.

In Chapter "Development of Congestion State Guiding System for University Cafeteria," Takafumi Doi, Hirotaka Ito, and Kenji Funahashi develop a congestion state guide system for a university cafeteria. This system confirms the congestion state of a cafeteria using iBeacon, and displays it on a mobile device in real time.

In Chapter "Analog Learning Neural Circuit with Switched Capacitor and the Design of Deep Learning Model," Masashi Kawaguchi, Naohiro Ishii, and Masayoshi Umeno, in a neural network field study, used analog electronic multiple and switched capacitor circuits. The connecting weights describe the input voltage.

In Chapter "Study on Category Classification of Conversation Document in Psychological Counseling with Machine Learning," Yasuo Ebara, Yuma Hayashida, Tomoya Uetsuji, and Koji Kotamada developed a system for visualizing the flow of conversation in counseling. They have implemented on the category classification method for text data of conversation document with SVM (support vector machine) as machine learning technique.

In Chapter "Improvement of "Multiple Sightseeing Spot Scheduling System"," Kazuya Murata and Takayuki Fujimoto develop "Multiple Sightseeing Spot Scheduling System" that enables a guide of new sightseeing in this research. In this paper, they consider an improvement plan for the application under development and implement the plan.

In Chapter "Advertising in the Webtoon of Cosmetics Brand—Focusing on 'tn' Youth Cosmetics Brands," Sieun Jeong, Hae-Kyung Chung, and Cheong-Ghil Kim suggest ways to improve the effects of advertisement setting and story structure, character setting and story structure, which can attract youth's interest by studying advertisement analysis of youth cosmetics brand Webtoon.

In Chapter "Testing Driven Development of Mobile Application Using Automatic Bug Management Systems," Mechelle Grace Zaragoza, Haeng-Kon Kim, In-Han Bae, and Jong-Hak Lee propose an approach to derive tests from the model of the mobile applications system as well as a diagram by using automatic bug management system. Using this technique, they can achieve more effective testing on hardware-related software areas.

In Chapter "Shape Recovery of Polyp from Endoscope Image Using Blood Vessel Information," Yuji Iwahori, Tomoya Suda, Kenji Funahashi, Hiroyasu Usami, Aili Wang, M. K. Bhuyan, and Kunio Kasugai aim to help medical doctor by proposing a new approach to estimate the size and 3D shape of polyp as a medical supporting system. This proposed approach uses blood vessel as a target with a known size to estimate the absolute size of polyp.

In Chapter "Design of Agent Development Framework for RoboCupRescue Simulation," Shunki Takami, Kazuo Takayanagi, Shivashish Jaishy, Nobuhiro Ito, and Kazunori Iwata design and implement an agent development framework for a RoboCup Rescue Simulation project that unifies the structure within the project to facilitate such technical exchange.

In Chapter "Mist Computing: Linking Cloudlet to Fogs," Minoru Uehara proposes placing a mist between cloudlets and fogs. The mist is a data center of cloudlets and a fog device. They describe the requirements and functions of mist computing.

In Chapter "Self-Recognition and Fault Awareness in OpenFlow Mech," Suguru Yasui and Minoru Uehara propose a computing system that adapts to environments and whose continued development will be made possible by replacing faulty and aging component in the system dynamically. They implement the two functions of self-recognition and fault awareness for a calculating unit in the system.

It is our sincere hope that this volume provides stimulation and inspiration, and that it will be used as a foundation for works to come.

July 2017   Takayuki Fujimoto
                                                                            Toyo University, Bunkyō, Japan

Contents

Contributors

Yoshiyuki Anan Base Division, Omron Software Co., Ltd., Shimogyo-ku, Kyoto, Japan

In-Han Bae Catholic University of Daegu, Gyeongsan, South Korea

M.K. Bhuyan Department of Electronics & Electrical Engineering IIT Guwahati, Guwahati, India

Hae-Kyung Chung Department of Moving Image Design, Konkuk University, Seoul, Republic of Korea

Takafumi Doi Nagoya Institute of Technology, Nagoya, Japan

Yasuo Ebara Academic Center for Computing and Media Studies, Kyoto University, Kyoto, Japan

Ziran Fan Graduate School of Information Sciences and Arts, Toyo University, Kawagoe-shi, Japan

Takayuki Fujimoto Graduate School of Information Sciences and Arts, Toyo University, Tokyo, Japan

Kenji Funahashi Department of Computer Science, Nagoya Institute of Technology, Nagoya, Japan

Yuma Hayashida Faculty of Engineering, Kyoto University, Kyoto, Japan

Naohiro Ishii Sugiyama Jogakuen University, Nagoya, Japan; Department of Information Science, Aichi Institute of Technology, Toyota, Yagusa-cho, Japan

Hirotaka Ito Nagoya Institute of Technology, Nagoya, Japan

Nobuhiro Ito Department of Information Science, Aichi Institute of Technology, Toyota, Aichi, Japan; Nagoya Institute of Technology, Nagoya, Japan

Yuji Iwahori Department of Computer Science, Chubu University, Kasugai, Japan

Kazunori Iwata Department of Business Administration, Aichi University, Nagoya, Aichi, Japan

Shivashish Jaishy Graduate School of Business Administration and Computer Science, Aichi Institute of Technology, Nagoya, Aichi, Japan

Sieun Jeong Department of Design, Konkuk University, Seoul, Republic of Korea

Yosuke Kanai Graduate School of Information Sciences and Arts, Toyo University, Kawagoe-shi, Japan

Masayoshi Kanoh Chukyo University, Showa-ku, Nagoya, Japan

Kunio Kasugai Department of Gastroenterology, Aichi Medical University, Nagakute, Aichi, Japan

Masashi Kawaguchi Department of Electrical & Electronic Engineering, Suzuka National College of Technology, Shiroko, Suzuka Mie, Japan

Cheong-Ghil Kim Department of Computer Science, Namseoul University, Cheonan, Republic of Korea; Catholic University of Daegu, Gyeongsan, South Korea

Haeng-Kon Kim Catholic University of Daegu, Gyeongsan, South Korea

Koji Koyamada Academic Center for Computing and Media Studies, Kyoto University, Kyoto, Japan

Jong-Hak Lee Catholic University of Daegu, Gyeongsan, South Korea

Kazuya Murata Graduate School of Engineering, Toyo University, Tokyo, Japan

Tsuyoshi Nakamura Nagoya Institute of Technology, Showa-ku, Nagoya, Japan

Toyoshiro Nakashima Department of Culture-Information Studies, Sugiyama Jogakuen University, Chikusa-ku, Nagoya, Aichi, Japan; Institute of Managerial Research, Aichi University, Nakamura-ku, Nagoya, Aichi, Japan

Kazuya Odagiri Sugiyama Jogakuen University, Nagoya, Japan

Daiki Sekiya Nagoya Institute of Technology, Showa-ku, Nagoya, Japan

Shogo Shimizu Sugiyama Jogakuen University, Nagoya, Japan

Tomoya Suda Department of Computer Science, Nagoya Institute of Technology, Showa-ku, Nagoya, Japan

Shunki Takami Graduate School of Business Administration and Computer Science, Aichi Institute of Technology, Nagoya, Aichi, Japan

Kazuo Takayanagi Graduate School of Business Administration and Computer Science, Aichi Institute of Technology, Nagoya, Aichi, Japan

Minoru Uehara Department of Information Sciences and Arts, Toyo University, Kawagoe, Japan

Tomoya Uetsuji Graduate School of Engineering, Kyoto University, Kyoto, Japan

Masayoshi Umeno Department of Electronic Engineering, Chubu University, Kasugai, Aichi, Japan

Hiroyasu Usami Department of Computer Science, Chubu University, Kasugai, Japan

Aili Wang Department of Communication Engineering, Harbin University of Science and Technology, Harbin, China

Koji Yamada Institute of Advanced Media Arts and Sciences, Ogaki, Gifu, Japan

Suguru Yasui Graduate School of Information Sciences and Arts, Toyo University, Kawagoe, Japan

Yukihiro Yoshida Nagoya Institute of Technology, Showa-ku, Nagoya, Japan

Mechelle Grace Zaragoza Catholic University of Daegu, Gyeongsan, South Korea

Proposed Framework Application for a Quality Mobile Application Measurement and Evaluation

Mechelle Grace Zaragoza and Haeng-Kon Kim

Abstract In order to keep the Capability Maturity Model levels four and higher, the quality of software development must be controlled by a quantification of the software development process as well as the product and the resources in the different phases. The quantification by means of software measurement needs a unified strategy, methodology or approach as one important prerequisite to guarantee the goals of quality assurance, improvement and controlled software management to be achieved. Nowadays, plenty of methods such as measurement frameworks, maturity model, goal-directed paradigms, process languages etc. exist to support this idea. However, the current approaches are based on a static view. This paper describes an Object-Oriented (OO) approach to a software measurement framework aimed at evaluating software products, software process, and resources. This approach includes the dynamic characteristics in software measurement, considering the behavior aspects of software metrics. The framework is described in principle based on some first practical applications.

Keywords Object-oriented approach · Mobile application · Software measurement and evaluation

1 Introduction

Software Measurement includes the phases of the modeling of the problem domain, the measurement, the presentation and analysis of the measurement values, and the evaluation of the modeled software with their relations shown in a simplified manner

M.G. Zaragoza · H.-K. Kim (✉)
Catholic University of Daegu, Gyeongsan, South Korea
e-mail: hangkon@cu.ac.kr

M.G. Zaragoza
e-mail: mechellezaragoza@gmail.com

© Springer International Publishing AG 2018
R. Lee (ed.), *Computational Science/Intelligence and Applied Informatics*,
Studies in Computational Intelligence 726, DOI 10.1007/978-3-319-63618-4_1

1

in Fig. 4. The most important thing to consider is the size and complexity of software systems are growing dramatically, in addition to which, the existence of automated tools leads to the generation of a huge number of test cases, the execution of which causes huge losses in cost and time. According to Rothermel et al., a product of about 20,000 lines of code requires seven weeks to run all its test cases [1].

As to Software Measurement, this enables the improvement and/or controlling of the measured software components. Software measurement exists more than twenty years. But, still, we can establish a lot of unsolved problems in this area. Some of these problems are

- the incompleteness of the chosen models.
- the restriction of the thresholds only for special evaluation in a special software environment,
- the weakness of the measurement atomization with metrics tools,
- the lack of metrics/measures validation,
- last but not least: the missing set of world-wide accepted measures and metrics including their units.

Software reliability engineering focuses on engineering mechanisms for quantitative evaluations of software reliability, the development of software, and the maintenance of software [2].

The problems of software measurement are also the main obstacles to the installation of metrics programs in an industrial environment. Software of program analysis is the process of analyzing the behavior of a computer program. Program analysis has a very widely application range, it provides support for compiler optimization, testing, debugging, verification and many other activities [3].

Hence, a measurement plan/framework is necessary which is based on the general experience of software measurement investigations. Today, numbers of methods such as measurement frameworks, maturity model, goal-directed paradigms, and process languages exist to support this idea. However, the current approaches are based on a static view. This paper describes an Object-Oriented (OO) approach to a software measurement framework aimed at evaluating software products, software process, and resources. This approach includes the dynamic characteristics in software measurement, considering the behavior aspects of software metrics. The framework is described in principle based on some first practical applications. The most substantial component in the endeavor to reach predictable performance and high capability software is the measurement of the software process and ensuring that process artifacts meet their specified quality requirements [4].

1.1   Explaining Software Measurement and Analysis

The initial phases of software product development require a careful analysis of the business to which the application is addressed. In particular, the analysis will identify information such as the goals and processes of the business.Software measurement is a process that is oriented methodical process that measures, evaluates, adjusts, and improves the software development process. [5]. Software Measurement is the process by which the organizations help to improve a measurement program [6]. Further, it introduces numerical, quantifiable engineering techniques into the world of software development [7]. Measurement is a map of the empirical world of the formal, relational world. Consequently, a measure is the number or symbol assigned to an entity by this mapping in order to characterize an attribute [8]. This assignment of numbers or symbols to any entity is made according to unambiguous rule. The rule of assignment can be any consistent rule excluding random assignment. An entity may be an object, such as a person or a software specification, or an event, such as an organization or a coded program [9].

Program analysis is the process of analyzing the behavior of a computer program. Program analysis has a very widely application range, it provides support for compiler optimization, testing, debugging, verification and many other activities [10].

1.2 What Is Software Architecture?

Software architecture is defined to be the rules, heuristics, and patterns governing: Partitioning the problem and the system to be built in discrete pieces Techniques used to create interfaces between these pieces. Techniques used to manage overall structure and flow. Techniques used to interface the system to its environment. Appropriate use of development and delivery approaches, techniques and tools.

1.3 Why Is Architecture Important?

The primary goal of software architecture is to define the non-functional requirements of a system and define the environment. The detailed design is followed by a definition of how to deliver the functional behavior within the architectural rules. Architecture is important because of the following:

- Controls complexity
- Enforces best practices
- Gives consistency and uniformity
- Increases predictability
- Enables re-use.

1.4 What Is an Object?

The object can be considered as a "thing", which can perform a number of relevant activities. All actions performed by the object determine the behavior of the object [11] (Fig. 1).

1.5 A Traditional System for Data Transmission Object-Oriented

Software In an object-oriented system, traditional parameters are usually applied to processes that involve class operations. The method is an integral part of an object that operates on the basis of data in response to the message and is defined in the context of the class declaration. Three traditional parameters are discussed as follows: cycloramic complexity, size (number of lines) and percent of the comments. The first approach to determine the set of measures aimed at the object was to focus on the main points of criticism of the designs of object-oriented design and the choice of measures for these areas. Parameters are supported by most publications and object-oriented tools. Measures to evaluate object-oriented concepts: methods, classes, and communication inheritance. The measures focus on the internal structure of objects, the external measurement of interaction between companies, measures the effectiveness of the algorithm and the use of machine resources and psychological measures to create an impact on the ability of the programmer and understand the changes and maintain the system [12].

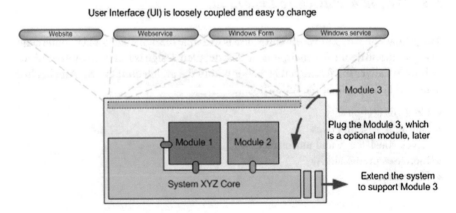

Fig. 1 Software architecture

2 Current Problem in Software Measurement

2.1 The Software Measurement Frameworks Situation

2.2 Object Oriented Testing

Software testing is a process of running a program or an application in search for the system's bugs. It can also be declared as the process of validation and verification software is applied to a software program or application or product. The testing of any software is an important means of measuring the software to define its quality. Problems Identified.

1. The introduction of new techniques technically and in accordance with the design of the existing techniques as OO Design.
2. The effectiveness of test production methods for all tests. In the context of the integration test: -Testing is the best way to ensure the quality of the product. It increased intensity in the level of integration.
3. Automatic test case generation
4. Prioritization of test cases- are automatically generated and one drawback is a test set for the. It prioritizes (TCP) that implies the prior planning of the execution of the execution of the test of relaying the effectiveness of the software testing activities in software process improvement [13]. Software measurement has a number of challenges, from both a theoretical and practical points of view. To cater challenges, a number of techniques that have been developed over the years and/or have been borrowed from other fields. The first thing to do is identifying, measuring and the features of software processes and products that are believed to be relevant and should be studied. Second is measuring these characteristics that are really valuable, with empirical validation of measures. An example for this is we need to show if and to what is the capacity of these characteristics influence to other characteristics of industrial interest, such as product reliability or simply the process cost.

   Lastly, the identification and assessment measures may not be valid in general. Goal oriented frameworks that have been defined for software measurement can be very useful [14]. A measurement framework should consider all these areas and their dependencies. This leads us to the TQM (Total Quality Management) approach. However, this is a highly complex and therefore an unsolved problem today, especially for new software development paradigms and techniques.

TQM requires considering all software components with all their (time-depended) relations-a typical functional approach. In the area of software development, the unsuitability of this approach is well-known and led to the paradigm-shifting towards object-orientation. Hence, we mean that the same change is necessary in the area of software measurement. For object-oriented software development, it is necessary to consider.

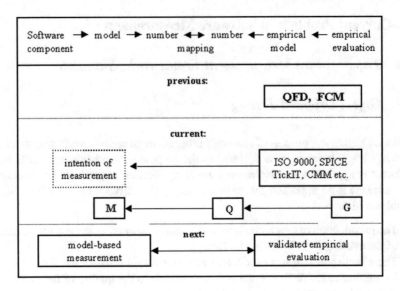

Fig. 2 Measurement framework situation

- the static structure(class hierarchy and their dependencies)
- The dynamic behavior (object collaboration and communication),
- the granularity of an OO system
- the special aspects of the OOA, OOD and OOP development phases.

On the other hand, we will take in our considerations the OO metrics experience presented in the next section (Fig. 2).

2.3 The OO Software Metrics Situation

We can establish the following general stages of OO metrics investigation given in the Fig. 3 [10].

However, the OO metrics are based on a 'functional' description such as [10].

- #classes = 'f(x)' = counting(all/selected classes),
- average number of methods per class = executing(#method/#classes),
- coupling between classes = counting(#called classes).

This is a typical static definition of an (OO) metric. On the other hand, metrics have also dynamic aspects such as the refinement during the software development, the tuning of the thresholds during the maintenance, the "communication" between metrics whereas estimation, and the controlling tasks of metrics. The modeling of these facilities is missing in the current OO software metrics area (Figs. 4 and 5).

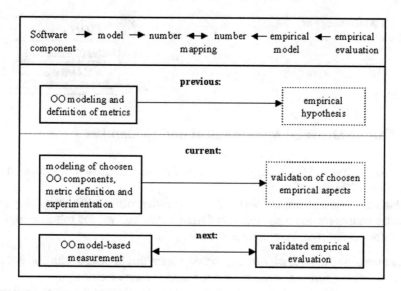

Fig. 3 OO software metrics situation

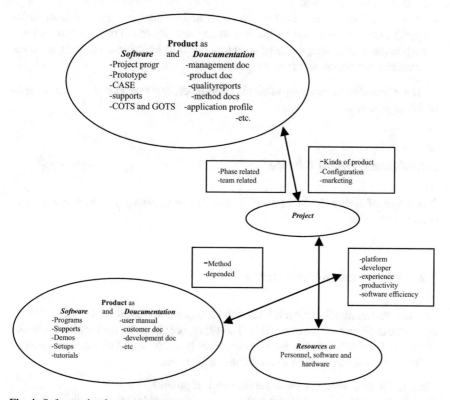

Fig. 4 Software development components

Fig. 5 CAME strategy of
measurement

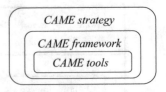

3   Object-Oriented Measurement and Evaluation

Framework: The CAME strategy is addressed in the background of measurement
intentions (Figs. 6 and 7).

Software metrics are a valuable entity in the entire software life cycle. They provide the measurement for the software development that includes software require-
ment documents, designs, programs and tests [15].

- Community. The necessity of a group or a team that is tasked with and has the
  knowledge of software measurement to install software metrics.
- Acceptance. The agreement of the (top) management to install a metrics program
  in the (IT) business area.
- Motivation. The production of measurement and evaluation results in a first metrics
  application that demonstrates the convincing benefits of the metrics application.
- Engagement. The acceptance of spending effort to implement the software measurement as a permanent metrics system [16].

The *attributes*: the metrics value characteristics, and *services*: the metrics application algorithms.

3.1   Measurement Choice

The choice of metrics includes the definition of object-oriented software metric as a
class/object with.

3.2   Measurement Adjustment

This measurement adjustment is related to the experience (expressed in values) of the
measured attributes for the evaluation. The adjustment includes the metrics validation
and the determination of the metrics algorithm based on the measurement strategy.

These are the steps of the measurement adjustment

Step 1. Determination of the scale type and (if possible) the unit
Step 2. Determination of the favorable values (thresholds) for the evaluation of the
         measurement component including their calibration

Software Metrics		
Process:	Size:	-number of components
		-size of components
		-volume of product lint
	Structure:	-life cycle model
		-management areas
		-organizational structure
	Complexity:	-dimensions of components
		-granularity of components
		-integrity of components
	Quality:	-maturity
		-certification
		-management (risks) level
Product:	Size:	-number of elements
		-size of product components
		-volume of versions
	Structure:	-design structure
		-implementation structure
		-architecture
	Complexity:	-psychological complexity
		-computational complexity
	Quality:	-functionality
		-reliability
		-efficiency
		-usability
		-maintainability
		-portability
Resources:	Size:	-number of platforms, teams etc.
		-size of standard software
		-volume of COTS
	Structure:	-team structure
		-computer network structure
		-system software structure
	Complexity:	-kinds of development cultures
		-parameters of platforms
		-dimensions of COTS
	Quality:	-personnel experience
		-product quality of COTS
		-hardware reliability

Fig. 6 Software metrics (class hierarchy) model

Fig. 7 The metrics class

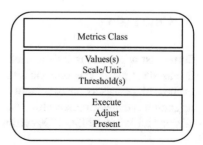

Metrics Class

Values(s)
Scale/Unit
Threshold(s)

Execute
Adjust
Present

Step 3. Tuning of the thresholds during the software development or maintenance
Step 4. Calibration of the scale depends on the improvement of the knowledge in the problem domain.

3.3 Measurement Migration

The migration step is addressed in the description of the behavior of a metric class, such as the "metrics tracing along the life cycle and metrics refinement along the software application". Thus, these aspects retain the dynamic characteristics that are essential for the persistent installation of metrics applications and require a metrics database or additional kinds of the metrics values background. Future research is directed at using the OO measurement framework to the UML-based development with the use of Java-based metrics class library.

3.4 Measurement Efficiency

This step contains the instrumentation or the automatization of the measurement process by tools. It needs to examine the algorithmic character of the software measurement and the prospect of the integration of tool-based "control cycles" in the software development process.

Tools supporting our framework are: CAME (Computer Assisted software Measurement and Evaluation) tools. In order to use:

1. (Combined) tool(s) must cover the entire measurement phases,
2. Tool(s) must consider the entire software life cycle,
3. CAME tools should keep the (ISO 9126) quality aspects among themselves.

The application of the CAME tools in the background of a metrics database is the first phase of measurement efficiency, and metrics class library the final OO framework installation [15, 16].

4 Conclusion

Some first applications are described in detail in applying CAME tools. However, in this short paper was only possible to indicate the principles and aspects of the framework phases as measurement choice, measurement adjustment, measurement migration and measurement efficiency. However, this approach clarifies the next steps after the initiatives of ISO 9000 certification, CMM evaluation and, on the other hand, the special metrics definition and analysis of small aspects. The further research effort is directed at using the OO measurement framework to the UML-based development method using a Java-based metrics class library.

Acknowledgements This work is supported by Catholic University of Daegu, Republic of Korea.

References

1. Saifan, A.A., Alsukhni, E., Alawneh, H., Sbaih, A.A.: Test case reduction using data mining technique. Int. J. Softw. Innov. (IJSI) **4**(4), 56–70 (2016)
2. Kang, M., Choi, O., Shin, J., Baik, J.: Improvement of software reliability estimation accuracy with consideration of failure removal effort. Int. J. Netw. Distrib. Comput. **1**(1), 25–36 (2013)
3. Chen, S., Sun, D., Miao, H.: The influence of alias and references escape on Java program analysis. In: Software Engineering Research, Management and Applications, pp. 99–111. Springer International Publishing (2015)
4. Unterkalmsteiner, M., Gorschek, T., Islam, A.M., Cheng, C.K., Permadi, R.B., Feldt, R.: Evaluation and measurement of software process improvement–a systematic literature review. IEEE Trans. Softw. Eng. **38**(2), 398–424 (2012)
5. Lee, M.C., Chang, T.: Software measurement and software metrics in software quality. Int. J. Softw. Eng. Appl. **7**(4), 15–34 (2013)
6. Sinha, B.K., Sinhal, A., Verma, B.: A software measurement using artificial neural network and support vector machine. Int. J. Softw. Eng. Appl. **4**(4), 41 (2013)
7. Easterbrook, S.: Empirical research methods for software engineering. In: Proceedings of the Twenty-Second IEEE/ACM International Conference on Automated Software Engineering, pp. 574–574. ACM(2007)
8. Farooq, S.U., Quadri, S.M.K., Ahmad, N.: Software measurements and metrics: role in effective software testing. Int. J. Eng. Sci. Technol. (IJEST) **3**(1), 671–680 (2011)
9. Timothy, B.: Introduction to Object-oriented Programming. Pearson Education India (2008)
10. Huang, R., Li, M., Li, Z.: Research of improving the quality of the object-oriented system. Int. J. Inf. Educ. Technol. **3**(4), 433 (2013)
11. Vinayak, A.: Research Issues in Object Oriented Software Testing, 18 Jan 2017
12. Jones, C.: Applied Software Measurement: Global Analysis of Productivity and Quality. McGraw-Hill Education Group, New York (2008)
13. Rawat, M.S., Mittal, A., Dubey, S.K.: Survey on impact of software metrics on software quality. Int. J. Adv. Comput. Sci. Appl. (IJACSA) **3**(1) (2012)
14. Elbert, C., Dumke, R., Bundschuh, M., Schmietendorf, A.: Best Practices in Softwre Measurement, How to Use Metrics to Improve Projet and Process Performance. Springer, Berlin (2005)
15. Dumke, R., Abran, A.: Software Measurement, 272 p. Springer Science & Business Media, Nov 11. Business & Economics
16. Kim, H.-K.: A Framework for Mobile Applications Quality Measurement and Evaluation

Proposal and Development of Artificial Personality(AP) Application Using the "Requesting" Mechanism

Yosuke Kanai and Takayuki Fujimoto

Abstract In recent years, people's interest in the artificial intelligence (AI) is increasing. Systems with AI technology are developed to be applied to various fields. The study model to create AI whose function simulates a human's mind and feelings and the developments are in progress. We assume that it is impossible to express "human-like qualities" by algorithm for autonomous AI with learning ability. Therefore, we develop the "Artifical Personality (AP)" that is a designed form of a personality without autonomous learning ability, a software to express more "human-like qualities" than AI by efforts for manual steps.

Keywords IT · AI · AP · OTAKU · MOE

1 Backgrounds and Motivations

In recent years, people's interest in the artificial intelligence (AI) is increasing. Systems with AI technology are developed to be applied to various fields [1]. AI is the computer system that is designed to play a role as a human substitute [2]. The study model to create AI whose function simulates a human's mind and feelings and the developments are in progress. For example, there are some products with AI system in widespread use: "Siri" equipped with iOS smart device and "Pepper", the personal AI robot that has the function of emotion-recognition. The most of them are the computers which aims to have human-like disposition. In other words, they have been developed along with the concept: "being close to humans" or "coexist with humans."

Y. Kanai (✉) · T. Fujimoto
Graduate School of Information Sciences and Arts, Toyo University, Kawagoe-shi, Japan
e-mail: kanaip@icloud.com

T. Fujimoto
e-mail: fujimoto@toyo.jp

© Springer International Publishing AG 2018
R. Lee (ed.), *Computational Science/Intelligence and Applied Informatics*,
Studies in Computational Intelligence 726, DOI 10.1007/978-3-319-63618-4_2

These days AI has become very popular and AI-like mechanisms are incorporated with various products and services.

However, the most of the products and services which are supposed to be with AI, do not always reflect the original concept: "human-like qualities." For example, they sometimes choose the best choice by calculation in a limited situation. At other times, they find the one-size-fits-all answer from the database in the situation in which one option needs to be selected [3]. At still other times, they are the basic AI chat program with the simple conversational function and the learning function to repeat Q and A with the user in case of its encountering unknown words or keywords. Most of them are named as AI and the study results and technologies in the field of AI are utilized for their development. However, when we actually use these products, the impressions are much different from what we call "human-like qualities." As a program or system, they emulate "human capabilities." Also, they have a human-like learning system or decision making system. Nevertheless, we cannot have the impression that they have "human-like" qualities [4]. We will not accept the product of AI system unless it is equipped with human-like qualities that enables us to imagine AI [5].

For example, AIBO was developed as a toy and it became popular widely. Most of the users are satisfied with "the gimmick like a real pet". For the artificial products, the expressive approach is supposed to be essential to produce the impression of vitality and human (animal)-like qualities, and it should be different from the greater sophistication or the increasingly complex of the AI program. However, from "The survey of Siri use" (imore 2015), only 4.74% users answered they use Siri many times every day and the 49.88% users answered they use Siri once in a month, or do not use it at all. The results show that AI is not used very much even if it is a precise artificial intelligence with excellence [6].

We assume that it is impossible to express "human-like qualities" by algorithm for autonomous AI with learning ability. Therefore, we develop the "Artifical Personality (AP)" that is a designed form of a personality without autonomous learning ability, a software to express more "human-like qualities" than AI by efforts for manual steps.

2   Purpose of the Study

From "The survey of Attitude about the AI" (BIGLOBE 2016), In response to the question: "Do you expect good service from AI? Or do you fear it?" 17% people answered "considerably expect good service from it" and the 37% people answered, "moderately expect good service from it". The rate of the answers of "expect good service from AI" is 54% in total, a majority in the survey.

Regarding another question: "What do you expect from AI?" 32% people answered, "to be a friend for communication," and 7% people answered "to be a partner/lover" The results show that people want the AI program to have communication skills. There are two purposes of this study.

First, our AP system aims to raise the users' interest in AP by invoking interactive communication between them and AP. Second, AP would emphasize the "human-like qualities" by making requests to the users for some actions.

In AP programming, making a response to the user's one-sided remark is not defined as an interactive communication [7]. In this study, we pay more attention to implement the AP fiction to talk to the user. We aim to make the user feel that AP has more "human-like qualities" by this function. Also, by the "requesting" function, AP can give a soft and tender impression, which is different from cold AI's the requests make the user want to be supportive and give the impression of human-like familiarity. We produce the new design to attract people to use AP. The mechanism of the system is much different from the other AI systems today. AI is the computer system that is designed to play a role as a human substitute. However, AP system proposed in this study requests the user for some actions. This system presents the new relationship with the user by leaving the impression of "the perfect AI system" and creating the familiar impression, "charming and open to the caring eyes" We aim to design that enables the user to have an attachment to AP system and get the user interested in the AP system. This "requesting" gimmick is fundamentally based on the Japanese OTAKU culture, "MOE"

"MOE" is the most important keyword in OTAKU culture. Even the people who have no interest in OTAKU culture, know the word, "MOE". The word "MOE" is used like an exclamation for the charming actions, statements and personality, whether they are real or fictitious. It is usually used by a male who loves the character of the fictitious beautiful girl. In OTAKU culture, "the game to experience pseudo-romance" has been popular among introverted males. The trend that the hard-core fans adore the characters blindly created the word "MOE". Acquiring the factor of "MOE", and targeting males that have interests in OTAKU culture as the users, we aim to design to build a strong attachment bond between AP and the users. Today, although a lot of products with AI system are sold in the market, as far as the author knows, there are no products with AP.

3 Relevant Studies

3.1 Siri

Siri is implemented in iOS smart devices. It is an application software to play a role of the user's secretary for iOS and MacOS' That uses natural language processing for recommendations and invoking the Web Service in response to the user's questions. The sound recognition of Siri is excellent, and the system is very accurate. However, from "The survey of Siri use" (imore 2015), only 4.74% users answered they use Siri many times every day and the 49.88% users answered they use Siri once in a month, or do not use it at all. The reasons for the answers are "no suitable occasions", "machinery voice", "I don't know what I can do with Siri ", and etc. The results show that AI is not used very much even if it is a precise artificial intelligence with excellence. For the users, the accurate system is not necessary because the user can use the voice input without Siri. we use SIRI for input through the voice, AI portray the "human-like" It will be more useful if get a result another answer from Google or web search. however, it is only reaction and joke if your input words have a keyword.

3.2 Pepper

Pepper is a robot with AI system, which is sold by Softbank. It is equipped with a function to estimate feelings from the human expression by a human's voice or face. The various gestures and answers to the human words are known well. It is necessary to carry out various registrations in advance. It seems that programming and creation of the usage by the owners are more important than the functions of Pepper itself. Pepper is a learning type robot with the concept: communication and interaction with humans. However, according to the questionnaire (2015: Mynavi) to ask if the respondents would like to have Pepper, 88.4% of the people answered "No." The reasons for the answer are "not cute", "not necessary", "too expensive", "scary", and etc. The results show that even though Pepper is known as the most artificial intelligence specialized in communication, it cannot actually communicate and interact with humans.

3.3 Aibo

Aibo is the animal type AI robot. Its first model "ERS-110" was released by Sony in 1999. It is said that more than 150,000 Aibos were sold in total before Sony's retreat from the robot business in 2006. Aibo can do quadruped walking and moves like a puppy. The users can interact with it as if it was a real pet. In addition, it is designed to grow up through the communication with the users. This robot was provided not to be a substitute for human works but to be the creature with loveable behaviors. AIBOs were the beloved robots not just because of the functions but also because of their charms. Some users held funerals or organ transplants in case of their breakdown.

4 Development Environment

In this study, we develop the system with "xcode" provided by Apple. The development language is "objective-C". The software is for iOS and we are going to release it in App Store after development.

5 The App Design

5.1 The Concept of the Developed App

The proposed app does not have autonomous learning ability. The app focuses on the communication with the user by applying reactions and personality especially for the targeted base of the users and the usages. There are many apps that focus on communication. However, when we have conversations with these apps, we feel awkwardness [8]. And we get the impressions of the machinery talks. This does not match with AI's concept that functions are aimed to imitate a human's mind and feelings. Therefore, our app was designed to enable conversations like a human's and evoke the user's feeling of attachment. We introduced "requesting" function to the prototype system, which enables the close relationship with the user. This function imitates a human's mood shifts. The AP implementation is realized by using a minimum.

Natural language processing system and a minimum mechanism with learning algorithm, and also by creating reactions with efforts for manual steps. This program is like a chatterbot, which has a simple AI chat system, and only has a minimum communication function. However, this system input the action from all user's comments and user's expectations. We express "human-like qualities" Specifically, our app targets men who love OTAKU cultures.

5.2 Prototype of App

Headings, or heads, are organizational devices that guide the reader through your paper. There are two types: component heads and text heads.

Today, there are a lot of items and services that have concept "humanity" and "coexist with humans" There are items and services developed especially for communication with humans. However, the most of the products and services which are supposed to be with AI, do not always reflect the original concept: "human-like qualities." We developed the app that includes new mechanism.

Answers and reactions to the comments

There are two patterns of reactions for the user's comments. AP choose an answer based on "the perfect match of the input word" or "the keyword extracted from the input word" This system has many answers to be chosen by perfect matching for the natural reactions. By registering answers to the user's remarks one-by-one manually, It resolves awkwardness of the AI chat programs. The learning type AI systems tend to generate awkward communication because they use frequently appearing words from the user's input in the past, and they simply put the word, input by the user as the answer of the questions, in the fixed phrases Our app has a response dictionary of many kinds of words, which enables the natural communication. Our study creates "human-like quality" by the perfect response dictionary. The accumulation of the registrations is in progress, asking 71 prospective users, so-called OTAKUs for advice and validation. Currently prototype system has about 1000 reaction words.

The requesting expression

We aimed to make the user think that AP has more "human-like qualities", which is unprecedented, by "requesting mechanism". In particular we pursued the satisfactory design to be a communication partner.

6   Prototype System

6.1   Main Mode

The following figures show the design image of the prototypic system. It is implemented by a dictionary with over 1000 reaction words now. The reaction is called by the word input by the user on a dictionary. This reaction is aimed to convey the impression of the humanlike lively comment based on humanlike thoughts. To make the impression, it is designed to generate more organic reactions, instead of the machinery and conventional reactions. Also, reactions are called with animation. This animation makes stronger feelings of the attachment. The screenshot of the app's main mode is indicated in Fig. 1.

A girl displayed in the center of the screen is the mascot (anthropomorphic image of AP) at the trial manufacture stage. The mascot's comment is displayed in the balloon on the top of the screen. The user can input the characters into a quadrangle column on the bottom of the screen. When the user presses the remark button, logs for remarks will be displayed in the quadrangle column. The comment on the user's

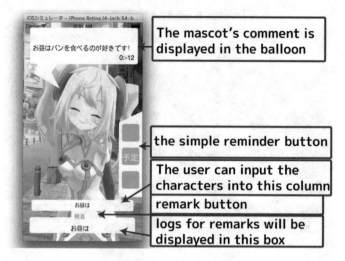

The mascot's comment is displayed in the balloon

the simple reminder button

The user can input the characters into this column

remark button

logs for remarks will be displayed in this box

Fig. 1 Main mode

remark will be displayed in the balloon on the top of the screen. At the same time, the mascot's image is animated as the reaction. This animation makes the impression with more humanity. This animation is a cutoff animation. This animation will play for subside situation if user input the word. There are about 10 kinds of animations.We express more human-like quality through these animations Fig. 2 show pattern of animations.

When AP system reaction for user, this app plays voice.

This voice used cute voice like an animation character because we targeting males that have interests in OTAKU culture as the users. They like these voices and we aim to design to build a strong attachment bond between AP and the users.

6.2 The Requesting Mode

"Requesting mode" is a unique function. AI is the computer system that is designed to play a role as a human substitute, The developed system requests something for the user. The AP and the user make requests each other. We present more human-like AP with the relationship Fig. 3 is the screen that will be displayed requesting mode started. Fig. 4 shows the screen that After fulfillment of a request.

In the prototype, the requesting mode is designed to open automatically by the time lapse. AP system requests something for the user, this means that the user is relied on by the system and the user is used by the AP system. Conventional AIs have the structure that the system provides proper response to the data input by the user, Our AP system have the structure that the system uses the user by making requests

Pattern of the animation

Fig. 2 Pattern of animations

Fig. 3 Requesting mode

besides of the conventional AI function. This may have the dilemma as the AI system to substitute the human's intellectual activity. However, by providing the two kinds of relationships: "The user uses the system" and "the system uses the user", This mechanism produces a pseudo-reliance between the user and the system. When a requesting mode opens, if the user do the action requested by the mascot, she will be pleased. By meeting the request, the point of reliance is increased. If this point is saved to some extent, AP's reaction will be provided in a more casual manner.

6.3 Simple Reminder Mode

Simple reminder mode is a simple function to set up a reminder to organize the schedule quickly.

Fig. 4 After fulfillment of a
request

Figure 5 is the screen that will be displayed when "the simple reminder button" is tapped. When the user presses the blue button displayed in the right of center on the screen, a simple reminder screen will be displayed and the user can set up a reminder by selecting alternatives. By choosing the time, place and action, a preview will be displayed in an orange quadrangle column on the bottom of the screen. When the user presses the "Confirm" button after choosing alternatives for all the items, the choices will be registered on a reminder. The user can come back to the main screen by pressing the "Back" button displayed on the bottom left of the screen. The user can confirm the input registration by pressing the red button displayed in the center of the screen again. Registrations for this simple reminder are limited for a daily schedule, and the user can organize the tasks of the day easily.

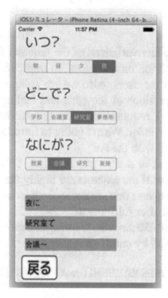 

Fig. 5 Simple reminder mode

7 Conclusion

In this study, we have experimentally developed the "Artificial Personality (AP)" with a concept: "the system to request something to the user." This is different from the other existent systems with AI. The user's actions for the system's requests are required by the mechanism. This feature makes AP's impression become closer to a human comparing with the learning AI system for which the user needs to provide his or her information for AI's better human-like qualities.

Therefore, we developed the "Artificial Personality (AP)" that is a designed form of a personality without autonomous learning ability, a software to express more "human-like qualities" with efforts for manual steps [9]. As the relevant studies showed, AIs are not used very much even if they are artificial intelligence with excellence precisely. The main reason for the phenomenon was "no suitable occasions". There are no suitable occasions for the users to use AIs on large devices, or any mobile devices like smartphones.

The application of this study does not just target to establish the new relationship between the user and the system by the requesting system but also focuses on making the user have feeling of attachment by the fact that user think he or she is relied on by the system. By incorporating the action, "requesting" into the application design, we aim to attract the users to use the application with emotional attachments [10].

8 Future Work

At the current development stage, there are few variations of the requests from the system, and it makes the application impression monotonous. To improve this, we are going to add other kinds of requests for the users. Also we are going to pursue "human-like qualities" by setting the impressions of thoughts and feelings of the AP finely. There are only two patterns of AP requests: creation or confirmation of the user's reminder, and creation of AP's reminder. We are going to introduce other practical functions to make the user involved with the AP.

For example, they can be a mail function or a note pad function. This system requires the user's response to its request, and if the requests are highly demanding, the user would feel tired. We will develop requests that are not too much challenging but give the user feeling of accomplishment after fulfillment.

3DCG model is used as a mascot of the AP, We gave the "human-like movements" to the mascot, which is a good point of 3DCG by cutoff animations.

Acknowledgements This work was supported by JSPS KAKENHI Grant Number 17K00730.

References

1. Syahanan, M.: From Singularity Artificial Intelligence to Super Intelligence, NTT
2. Baba, N., Yamada, M.: Foundations of Artificial Intelligence
3. Taniguch, T.: Learning Artificial Intelligence by Illust, Kodansya
4. Kurzweil, R.: The Singularity is Near: When Humans Transcend Biology
5. Sakuma, T., Kato, S.: A Ball Game Tipe Interaction Based on Anotheropomorphic Agents Reflecting User's Tendency of Giving Reward
6. Nakathuji, M., Fujiwara, Y., Uchiyama, T., Toda, H.: Collaborative Filtering by Analyzung Dynamic User Interests Modeled by Tax-onomy
7. Shibata, M., Tomiura, Y., Nishiguti, T.: Method for Selecting Appropriate Sentence from Documents on the WWW for the Open-ended Conversation Dialog System
8. Kanai, Y., Fujimoto, T.: Development of an Idol Entertainment Application with Focus on "Hagashi" Act
9. Fujimoto, T.: What is information design and what is not information design
10. Moribe, Y.: Information Design and its Directionality

Load Experiment of the vDACS Scheme in Case of the 300 Simultaneous Connection

Kazuya Odagiri, Shogo Shimizu and Naohiro Ishii

Abstract In the current Internet system, there are many problems using anonymity of the network communication such as personal information leaks and crimes using the Internet system. This is why TCP/IP protocol used in Internet system does not have the user identification information on the communication data, and it is difficult to supervise the user performing the above acts immediately. As a study for solving the above problem, there is the study of Policy Based Network Management (PBNM). This is the scheme for managing a whole Local Area Network (LAN) through communication control for every user. In this PBNM, two types of schemes exist. The first is the scheme for managing the whole LAN by locating the communication control mechanisms on the path between network servers and clients. The second is the scheme of managing the whole LAN by locating the communication control mechanisms on clients. As the second scheme, we have studied theoretically about the Destination Addressing Control System (DACS) Scheme. By applying this DACS Scheme to Internet system management, we will realize the policy-based Internet system management. In this paper, as the progression phase for the last goal, we perform the load experiment of the cloud type virtual PBNM named the vDACS Scheme, which can be used by plural organizations, for applications to the small and medium size scale organization. The number of clients used in an experiment is 300.

1 Introduction

In the current Internet system, there are many problems using anonymity of the network communication such as personal information leaks and crimes using the Internet system. The news of the information leak in the big company is sometimes

K. Odagiri (✉) · S. Shimizu · N. Ishii
Sugiyama Jogakuen University, Nagoya, Japan
e-mail: kazuodagiri@yahoo.co.jp

S. Shimizu
e-mail: shogo.shimizu@gakushuin.ac.jp

N. Ishii
e-mail: ishii@aitech.ac.jp

© Springer International Publishing AG 2018
R. Lee (ed.), *Computational Science/Intelligence and Applied Informatics*,
Studies in Computational Intelligence 726, DOI 10.1007/978-3-319-63618-4_3

reported through the mass media. Because TCP/IP protocol used in Internet system does not have the user identification information on the communication data, it is difficult to supervise the user performing the above acts immediately. As studies and technologies for managing Internet system realized on TCP/IP protocol, those such as Domain Name System (DNS), Routing protocol, Fire Wall (F/W) and Network address port translation (NAPT)/network address translation (NAT) are listed. Except these studies, various studies are performed elsewhere. However, they are the studies for managing the specific part of the Internet system, and have no purpose of solving the above problems.

As a study for solving the problems, Policy Based Network Management (PBNM) [2] exists. The PBNM is a scheme for managing a whole Local Area Network (LAN) through communication control every user, and cannot be applied to the Internet system. This PBNM is often used in a scene of campus network management. In a campus network, network management is quite complicated. Because a computer management section manages only a small portion of the wide needs of the campus network, there are some user support problems. For example, when mail boxes on one server are divided and relocated to some different server machines, it is necessary for some users to update a client machine's setups. Most of computer network users in a campus are students. Because students do not check frequently their e-mail, it is hard work to make them aware of the settings update. This administrative operation is executed by means of web pages and/or posters. For the system administrator, individual technical support is a stiff part of the network management. Because the PBNM manages a whole LAN, it is easy to solve this kind of problem. In addition, for the problem such as personal information leak, the PBNM can manage a whole LAN by making anonymous communication non-anonymous. As the result, it becomes possible to identify the user who steals personal information and commits a crime swiftly and easily. Therefore, by applying the PBNM, we will study about the policy-based Internet system management.

In the existing PBNM, there are two types of schemes. The first is the scheme of managing the whole LAN by locating the communication control mechanisms on the path between network servers and clients. The second is the scheme of managing the whole LAN by locating the communication control mechanisms on clients. It is difficult to apply the first scheme to Internet system management practically, because the communication control mechanism needs to be located on the path between network servers and clients without exception. Because the second scheme locates the communication control mechanisms as the software on each client, it becomes possible to apply the second scheme to Internet system management by devising the installing mechanism so that users can install the software to the client easily.

As the second scheme, we have studied theoretically about the Destination Addressing Control System (DACS) Scheme. As the works on the DACS Scheme, we showed the basic principle of the DACS Scheme, and security function [14]. After that, we implemented a DACS System to realize a concept of the DACS Scheme. By applying this DACS Scheme to Internet system, we will realize the policy-based Internet system management. Then, the Wide Area DACS system (wDACS system) [15] to use it in one organization was showed as the second phase for the last goal.

As the first step of the second phase, we showed the concept of the cloud type virtual PBNM, which could be used by plural organizations [16]. In this paper, as the progression phase of the third phase for the last goal, we perform the load experiment to confirm the possibility of the cloud type virtual PBNM for the use in plural organizations. In Sect. 2, motivation and related research for this study are described. In Sect. 2, the existing DACS Scheme and wDACS Scheme is described. In Sect. 4 the proposed scheme and load experiment results are described.

2   Motivation and Related Research

In the current Internet system, problems using anonymity of the network communication such as personal information leak and crimes using the Internet system occur. Because TCP/IP protocol used in Internet system does not have the user identification information on the communication data, it is difficult to supervise the user performing the above acts immediately.

As studies and technologies for Internet system management to be comprises of TCP/IP [1], many technologies are studied. For examples, Domain name system (DNS), Routing protocol such as Interior gateway protocol (IGP) such as Routing information protocol (RIP) and Open shortest path first (OSPF), Fire Wall (F/W), Network address translation (NAT)/Network address port translation (NAPT), Load balancing, Virtual private network (VPN), Public key infrastructure (PKI), Server virtualization. Except these studies, various studies are performed elsewhere. However, they are for managing the specific part of the Internet system, and have no purpose of solving the above problems.

As a study for solving the above problem, the study area about PBNM exists. This is a scheme of managing a whole LAN through communication control every user. Because this PBNM manages a whole LAN by making anonymous communication non-anonymous, it becomes possible to identify the user who steals personal information and commits a crime swiftly and easily. Therefore, by applying this policy-based thinking, we study about the policy-based Internet system management.

In policy-based network management, there are two types of schemes. The first scheme is the scheme described in Fig. 1. The standardization of this scheme is performed in various organizations. In IETF, a framework of PBNM [2] was established. Standards about each element constituting this framework are as follows. As a model of control information stored in the server called Policy Repository, Policy Core Information model (PCIM) [3] was established. After it, PCMIe [4] was established by extending the PCIM. To describe them in the form of Lightweight Directory Access Protocol (LDAP), Policy Core LDAP Schema (PCLS) [5] was established. As a protocol to distribute the control information stored in Policy Repository or decision result from the PDP to the PEP, Common Open Policy Service (COPS) [6] was established. Based on the difference in distribution method, COPS usage for RSVP (COPS-RSVP) [7] and COPS usage for Provisioning (COPS-PR) [8] were established. RSVP is an abbreviation for Resource Reservation Protocol.

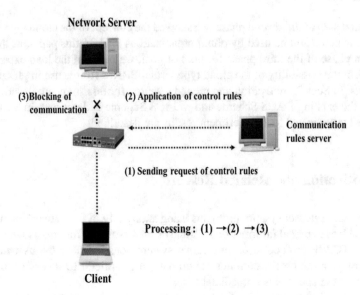

Network Server

(3)Blocking of   (2) Application of control rules
    communication   ✕

                                                        Communication
                                                        rules server

                        (1) Sending request of control rules

                        Processing: (1) →(2) →(3)

Client

Fig. 1 Principle in first scheme

The COPS-RSVP is the method as follows. After the PEP having detected the communication from a user or a client application, the PDP makes a judgmental decision for it. The decision is sent and applied to the PEP, and the PEP adds the control to it. The COPS-PR is the method of distributing the control information or decision result to the PEP before accepting the communication.

Next, in DMTF, a framework of PBNM called Directory-enabled Network (DEN) was established. Like the IETF framework, control information is stored in the server storing control information called Policy Server, which is built by using the directory service such as LDAP [9], and is distributed to network servers and networking equipment such as switch and router. As the result, the whole LAN is managed. The model of control information used in DEN is called Common Information Model (CIM), the schema of the CIM (CIM Schema Version 2.30.0) [11] was opened. The CIM was extended to support the DEN [10], and was incorporated in the framework of DEN.

In addition, Resource and Admission Control Subsystem (RACS) [12] was established in Telecoms and Internet converged Services and protocols for Advanced Network (TISPAN) of European Telecommunications Standards Institute (ETSI), and Resource and Admission Control Functions (RACF) was established in International Telecommunication Union Telecommunication Standardization Sector (ITU-T) [13].

However, all the frameworks explained above are based on the principle shown in Fig. 1. As problems of these frameworks, two points are presented as follows. Essential principle is described in Fig. 2. To be concrete, in the point called PDP (Policy Decision Point), judgment such as permission and non-permission for communication pass is performed based on policy information. The judgment is notified and

Fig. 2 Essential principle

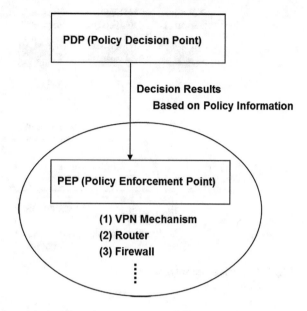

transmitted to the point called the PEP, which is the mechanism such as VPN mechanism, router and Fire Wall located on the network path among hosts such as servers and clients. Based on that judgment, the control is added for the communication that is going to pass by.

The principle of the second scheme is described in Fig. 3. By locating the communication control mechanisms on the clients, the whole LAN is managed. Because this scheme controls the network communications on each client, the processing load is low. However, because the communication control mechanisms need to be located on each client, the work load becomes heavy.

When it is thought that Internet system is managed by using these two schemes, it is difficult to apply the first scheme to Internet system management practically. This is why the communication control mechanism needs to be located on the path between network servers and clients without exception. On the other hand, the second scheme locates the communication controls mechanisms on each client. That is, the software for communication control is installed on each client. So, by devising the installing mechanism letting users install software to the client easily, it becomes possible to apply the second scheme to Internet system management. As a first step for the last goal, we showed the Wide Area DACS system (wDACS) system [15]. This system manages a wide area network, which one organization manages. Therefore, it is impossible for plural organizations to use this system. Then, as the first step of the second phase, we showed the concept of the cloud type virtual PBNM, which could be used by plural organizations in this paper.

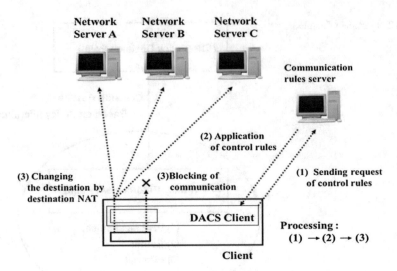

Fig. 3 Principle in second scheme

3 Existing DACS Scheme and wDACS System

In this section, the content of the DACS Scheme which is the study of the phase 1 is described.

3.1 Basic Principle of the DACS Scheme

Figure 4 shows the basic principle of the network services by the DACS Scheme. At the timing of the (a) or (b) as shown in the following, the DACS rules (rules defined by the user unit) are distributed from the DACS Server to the DACS Client.

(a) At the time of a user logging in the client.
(b) At the time of a delivery indication from the system administrator.

According to the distributed DACS rules, the DACS Client performs (1) or (2) operation as shown in the following. Then, communication control of the client is performed for every login user.

(1) Destination information on IP Packet, which is sent from application program, is changed.
(2) IP Packet from the client, which is sent from the application program to the outside of the client, is blocked.

An example of the case (1) is shown in Fig. 4. In Fig. 4, the system administrator can distribute a communication of the login user to the specified server among

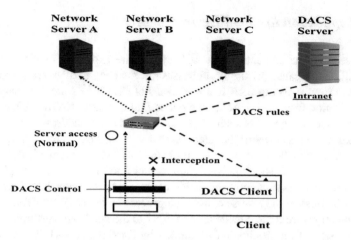

Fig. 4 Basic principle of the DACS scheme

Fig. 5 Layer setting of the DACS scheme

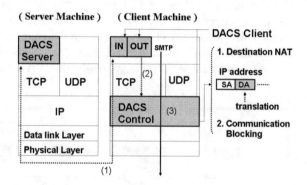

servers A, B or C. Moreover, the case (2) is described. For example, when the system administrator wants to forbid an user to use MUA (Mail User Agent), it will be performed by blocking IP Packet with the specific destination information.

In order to realize the DACS Scheme, the operation is done by a DACS Protocol as shown in Fig. 5. As shown by (1) in Fig. 5, the distribution of the DACS rules is performed on communication between the DACS Server and the DACS Client, which is arranged at the application layer. The application of the DACS rules to the DACS Control is shown by (2) in Fig. 5.

The steady communication control, such as a modification of the destination information or the communication blocking is performed at the network layer as shown by (3) in Fig. 5.

3.2 Application to Cloud Environment

In this section, the contents of wDACS system are explained in Fig. 6. First, as preconditions, because private IP addresses are assigned to all servers and clients existing in from LAN1 to LAN n, mechanisms of NAT/NAPT are necessary for the communication from each LAN to the outside. In this case, NAT/NAPT is located on the entrance of the LAN such as (1), and the private IP address is converted to the global IP address towards the direction of the arrow. Next, because the private IP addresses are set on the servers and clients in the LAN, other communications except those converted by Destination NAT cannot enter into the LAN. But, responses for the communications sent form the inside of the LAN can enter into the inside of the LAN because of the reverse conversion process by the NAT/NAPT. In addition, communications from the outside of the LAN1 to the inside are performed through the conversion of the destination IP address by Destination NAT. To be concrete, the global IP address at the same of the outside interface of the router is changed to the private IP address of each server. From here, system configuration of each LAN is described. First, the DACS Server and the authentication server are located on the DMZ on the LAN1 such as (4). On the entrance of the LAN1, NAT/NAPT and destination NAT exists such as (1) and (2). Because only the DACS Server and network servers are set as the target destination, the authentication server cannot be

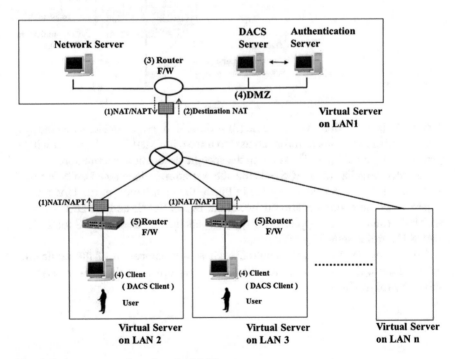

Fig. 6 Basic system configuration of wDACS system

accessed from the outside of the LAN1. In the LANs form LAN 2 to LAN n, clients managed by the wDACS system exist, and NAT/NAPT is located on the entrance of each LAN such as (1). Then, F/W such as (3) or (5) exists behind or with NAT/NAPT in all LANs.

4   The Cloud Type Virtual PBNM for the Common Use Between Plural Organizations

In this section, after the concept and implementation of the proposed scheme were described, functional evaluation results are described.

4.1   Concept of the Cloud Type Virtual PBNM for the Common Use Between Plural Organizations

In Fig. 7 which is described in [16], the proposed concept is shown. Because the existing wDACS Scheme realized the PBNM control with the software called the DACS Server and the DACS client, other mechanism was not needed. By this point, application to the cloud environment was easy.

The proposed scheme in this paper realizes the common usage by plural organizations by adding the following elements to realize the common usage by plural organizations: user identification of the plural organizations, management of the policy information of the plural organizations, application of the PKI for code communication in the Internet, Redundant configuration of the DACS Server (policy information server), load balancing configuration of the DACS Server, installation function of DACS Client by way of the Internet.

4.2   Implementation of the Basic Function in the Cloud Type Virtual PBNM for the Common Usage Between Plural Organizations

In the past study [14], the DACS Client was operated on the windows operation system (Windows OS). It was because there were many cases that the Windows OS was used for as the OS of the client. However, the Linux operating system (Linux OS) had enough functions to be used as the client recently, too. In addition, it was thought that the case used in the clients in the future came out recently. Therefore, to prove the possibility of the DACS Scheme on the Linux OS, the basic function of the DACS Client was implemented in this study. The basic functions of the DACS Server

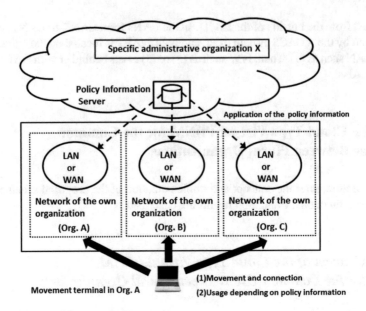

Fig. 7 Concept of the proposed scheme

and DACS Client were implemented by JAVA language. From here, it is described about the order of the process in the DACS Client and DACS Server as follows.

(Processes in the DACS Client)

(p1) The information acquisition from Cent OS

(p2) Transmission from the DACS Client to the DACS

(p3) The information transmission from the DACS Client to

(p4) The reception of the DACS rules from the DACS Server

(p5) Application of the DACS rules of the DACS Control

(Processes in the DACS Server)

(p1) The information reception from the DACS Client

(p2) Connection to the database

(p3) Inquiry of the Database

(p4) Transmission of the DACS rules to the DACS Client

4.3 Results of the Functional Evaluation

In this section, the results of the functional evaluation for the implementation system are described in Fig. 8.

In Fig. 9, the setting situation of the DACS rules is described in Fig. 9. This DACS rules is the rule to change a Web server for the access. The delivery of the DACS rules is between the DACS SV and the DACS CL encrypted by using SSL.

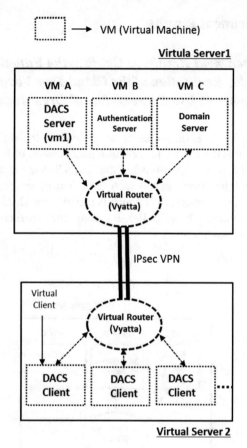

Fig. 8 Prototype system

```
<?XML version="1.0" encoding="uft8"?>
<direct>
  <rule priority="0" table="nat" ipv="ipv4" chain="PREROUTING_direct">
-d 192.168.1.10:80 -j DNAT --to 192.168.1.12:80</rule>
</direct>
```

Fig. 9 Setting situation of the DACS rules on the DACS CL

By this DACS rules, the next operation was realized. When the user accessed the Web Server with the IP address of 192.168.1.10, the Web Server with the IP address of 192.168.1.12 was accessed actually. As for this communication result, the communication log on each Web server was confirmed by viewing.

5   Load Experiment Results

5.1   Load Experiment Results to Confirm the Function of the Software for Realization of the Cloud Type Virtual PBNM for the Common Use Between Plural Organizations

In this section, the load experiment results are described. In the Fig. 10, the experimental environment is described. This environment consists of four virtual servers. In the virtual server 1, servers group such as the DACS Server and user authentication server is stored. In other virtual severs such as the virtual server 2, virtual server 3 and virtual server 4, the virtual client which is installed the DACS Client is stored. The number of the virtual clients is 100. By using this experimental environment, the load experiment was executed.

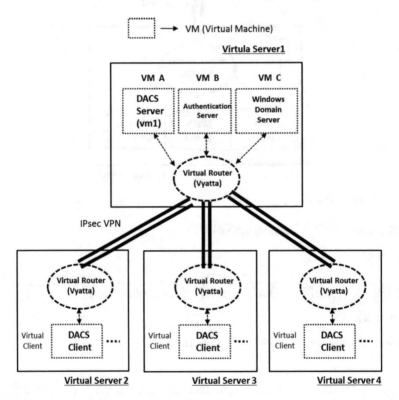

Fig. 10 Experimental environment

5.2   Load Experiment Results for Applications to the Small and Medium Size Scale Organization

(Experimental results by 100 clients)

The average of the results of the measurement for ten times was 263.2 MHz.

(Experimental results by 200 clients)

The average of the CPU consumption was 540.1 MHz. The value of around 540 MHz is the CPU load when the 200 clients are connected simultaneously.

After these experiments, the load experiments by 300 clients were performed. The experimental environment is described. The experimental environment is as previous experiment environment. The results are described as follows.

(Case of 10 simultaneous connections at the DACS SV)

In the first experiment, the number of the simultaneous connection for the DACS SV was set to 10 on this occasion. The experimental results are described in Fig. 11.

In this Figure, the average of the results of the measurement for ten times was 732.8 MHz.

(Case of 20 simultaneous connections at the DACS SV)

In the second experiment, the number of the simultaneous connection for the DACS SV was set to 20 on this occasion. The experimental results are described in Fig. 12.

In this Figure, the average of the results of the measurement for ten times was 750.1 MHz.

(Case of 30 simultaneous connections at the DACS SV)

In the third experiment, the number of the simultaneous connection for the DACS SV was set to 20 on this occasion. The experimental results are described in Fig. 13.

In this Figure, the average of the results of the measurement for ten times was 748.8 MHz.

	Practice Time	CPU Consumption
1	17:00	724
2	17:15	724
3	17:30	745
4	17:45	730
5	18:00	749
6	18:15	737
7	18:30	725
8	18:45	754
9	19:00	717
10	19:15	723

Fig. 11 Experimental results (1)

	Practice Time	CPU Consumption
1	20:45	766
2	21:00	748
3	21:15	763
4	21:30	777
5	21:45	785
6	22:00	716
7	22:15	753
8	22:30	721
9	22:45	744
10	23:00	728

Fig. 12 Experimental results (2)

	Practice Time	CPU Consumption
1	0:15	758
2	0:30	767
3	0:45	754
4	1:00	743
5	1:15	742
6	1:30	756
7	1:45	730
8	2:00	752
9	2:15	720
10	2:30	766

Fig. 13 Experimental results (3)

As the results of these experiments, the CPU consumption becomes approximately constant in the case of the number of the simultaneous connection from 20 to 30. The value of around 750 MHz is the CPU load when the 300 clients are connected simultaneously.

6   Conclusion

In this paper, we performed the load experiment of the cloud type virtual PBNM, which can be used by plural organizations. In this experiment, the 300 virtual clients with Linux OS are used, and the communications between the DACS SV and the DACS CL are encrypted. As the result, the average of CPU consumption was 750.1 MHz. When the number of the simultaneous connection for the DACS SV was set to 30 on this occasion, the average of CPU consumption was 748.8 MHz. These two values are 2.8 times as large as the value in case of the 10 simultaneous connections.

As a future work, we are going to perform more load experiments in the form of increasing the number of the virtual client and the number of the simultaneous connection for the DACS SV.

Acknowledgements This work was supported by the research grant of KDDI Foundation. We express our gratitude.

References

1. Cerf, V., Kahn, E.: A protocol for packet network interconnection. IEEE Trans. Commun. COM-22, 637–648 (1974)
2. Yavatkar, R., Pendarakis, D., Guerin, R.: A framework for policy-based admission control. IETF RFC 2753 (2000)
3. Moore, B., at el.: Policy Core Information Model–Version 1 Specification. IETF RFC 3060 (2001)
4. Moore, B.: Policy core information model (PCIM) extensions. IETF 3460 (2003)
5. Strassner, J., Moore, B., Moats, R., Ellesson, E.: Policy core lightweight directory access protocol (LDAP) schema. IETF RFC 3703 (2004)
6. Durham, D., et al.: The COPS (Common Open Policy Service) protocol. IETF RFC 2748 (2000)
7. Herzog, S., at el.: COPS usage for RSVP. IETF RFC 2749 (2000)
8. Chan, K., et al.: COPS usage for policy provisioning (COPS-PR). IETF RFC 3084 (2001)
9. CIM Core Model V2.5 LDAP Mapping Specification (2002)
10. Wahl, M., Howes, T., Kille, S.: Lightweight Directory Access Protocol (v3). IETF RFC 2251 (1997)
11. CIM Schema: Version 2.30.0 (2011)
12. Etsi, E.S.: 282 003: Telecoms and Internet converged Services and protocols for Advanced Network (TISPAN); Resource and Admission Control Subsystem (RACS). Functional, Architecture (2006)
13. ETSI ETSI ES 283 026: Telecommunications and Internet Converged Services and Protocols for Advanced Networking (TISPAN); Resource and Admission Control; Protocol for QoS reservation information exchange between the Service Policy Decision Function (SPDF) and the Access-Resource and Admission Control Function (A-RACF) in the Resource and Protocol specification, April 2006
14. Odagiri, K., Yaegashi, R., Tadauchi, M., Ishii, N.: Secure DACS scheme. J. Netw. Comput. Appl. 31(4), 851–861 (2008)
15. Odagiri, K., Shimizu, S., Takizawa, M., Ishii, N.: Theoretical suggestion of policy-based wide area network management system (wDACS system part-I). Int. J. Netw. Distrib. Comput. (IJNDC) 1(4), 260–269 (2013)
16. Odagiri, K., Shimizu, S., Ishii, N., Takizawa, M.: Suggestion of the cloud type virtual policy based network management scheme for the common use between plural organizations. In: Proceedings of International Conference on International Conference on Network-Based Information Systems (NBiS-2015), pp. 180-186, September 2015

Hearing-Dog Robot to Wake People Up Using its Bumping Action

Yukihiro Yoshida, Daiki Sekiya, Tsuyoshi Nakamura, Masayoshi Kanoh and Koji Yamada

Abstract We propose a robot inspired by the behavior of hearing dogs. Hearing dog is a guide dog for deaf people by notifying them of some life sounds such as fire alarms, doorbells, wake-up calls and so on. Hearing dogs use their touching behavior to communicate deaf people. This is a kind of haptic communications. The proposed robot also uses physical contact to communicate the people like hearing dog's touching behavior. The robot can approach and bump the people to notify of some life sounds happened around the people. This paper utilized the robot as wake-up call. We conducted an experiment to evaluate the usefulness of the robot to wake up sleeping people. This paper reports the experimental results and discusses usefulness of the robot's wake-up call.

1 Introduction

Sleeping situation is risky for deaf people because some alarms are often notified by sounds. According to this some of deaf people use smartphones or wearable devices which deliver vibrations as an alarm. Smartphones and wearable devices are useful for deaf people, but such equipment do not always work as expected. For example, deaf people set their smartphones under their pillows, but once their head

Y. Yoshida · D. Sekiya · T. Nakamura (✉)
Nagoya Institute of Technology, Gokiso-cho, Showa-ku, Nagoya, Japan
e-mail: tnaka@ai.nitech.ac.jp

Y. Yoshida
e-mail: yoshida@ai.nitech.ac.jp

D. Sekiya
e-mail: sekiya@ai.nitech.ac.jp

M. Kanoh
Chukyo University, 101-2 Yagoto Honmachi, Showa-ku, Nagoya, Japan
e-mail: mkanoh@sist.chukyo-u.ac.jp

K. Yamada
Institute of Advanced Media Arts and Sciences, 4-1-7 Kagano, Ogaki, Gifu, Japan
e-mail: k-yamada@iamas.ac.jp

© Springer International Publishing AG 2018   41
R. Lee (ed.), *Computational Science/Intelligence and Applied Informatics*,
Studies in Computational Intelligence 726, DOI 10.1007/978-3-319-63618-4_4

leaves the pillow, the smartphone cannot notify alarms. Wearable devices such as a smart watch, their vibrations can directly reach the user's body, but some people feel uncomfortable wearing such devices while sleeping. Dead battery is also a problem.

On the other hand, hearing dogs [1–3] can assist deaf people by alerting their handlers to some life sounds. For example, the life sounds are wake-up calls, doorbells, incoming emails, oven buzzers, fire alarms, crying babies, and so on. The dogs use physical contact, which is an effective way to communicate deaf people. Physical contact is a haptic communication which is a nonverbal and nonvisual communication. Such communication can offer people and animals to communicate and interact. Meanwhile there are fewer hearing dogs in comparison with the guide dogs for visually impaired people. This is because it is difficult to train hearing dogs to have high capability for assistance. Addition to this, hearing dogs are living animals and cannot always share living environments with humans.

We propose a robot which can alert deaf people to sounds related to their life. The robot was modeled on the touch behavior of hearing dogs. Hearing dogs can walk and touch people, and their touch aims to notify people of life sounds. The proposed robot can move using its own wheels and bump chairs or beds that people are sitting or sleeping on. The bumping action is inspired by the touch behavior of hearing dogs. Unlike living animals, robots have fewer problems to share living environment with humans.

Our study aims to develop a robot with a functionality of hearing-dogs. Hearing dogs have various kinds of functionality, and we focused on notification of life sounds. In order to implement the notification of life sounds, we designed a behavior model of the robot. According to the behavior model, the robot recognizes the sounds, searches for the user, and notifies the user of the sounds. In this paper the behavior model was implemented on Turtlebot, which has wheels to move and bumper sensors to recognize bumping the user. Turtlebot is suitable to develop and test a concept model of the hearing-dog robot.

This paper reports the concept model of the hearing-dog robot and discusses an optimal behavior design of the robot to wake sleeping people up. The robot is larger and heavier than smartphones and wearable devices, its physical contact can be much stronger than that of the vibration of those devices. In addition, a user does not need to wear the robot, and it can move and recharge its battery autonomously. We conducted an experiment to evaluate usefulness of the robot bumping to wake sleeping people up.

2 Related Works

There are some works have suggested that human-dog interaction can be useful for designing social behaviors of robots interacting with humans [4, 5]. Koay et al. [6] proposed a prototype of a hearing robot, which hires visual communication signals to communicate with people. That is to say, their work focused on the visual behavior of hearing dogs.

Furuhashi et al. [7] investigated the psychological effects of physical contact of a robot. Their experiment recruited participants with hearing difficulties to evaluate the effects, but the experiment wasn't conducted to wake sleeping people up. Sekiya et al. [8] conducted an experiment to wake sleeping people up using bumping behavior of a robot. Their work is most related to this paper, but their work recruited only ten participants. Therefore it is difficult that the experimental result can provide high reliability. On the other hand, we recruited 16 participants to improve the reliability of the experiment as described in this paper.

Robots have real bodies and physical existence in the real world, thus differing from computer-graphic virtual agents [9–12]. There had been some approaches to use physical contact to communicate between humans and robots. Nakagawa et al. [13] reported some changes in the motivation of humans, depending on whether physical touch was involved when a robot required humans to perform some tasks. Their result suggests effectiveness of physical touch between humans and robots. Chen et al. [14] proposed a nurse robot. Their study investigated the human reaction to physical touch by the robot when touch was accompanied by speech. Although our proposed robot also uses physical bumping for human-robot interaction, the robots that are reported in the previous studies differ from our robot in terms of both the aim and the approach.

3   Robot Configuration

The robot system configuration and the appearance of the robot are illustrated in Figs. 1 and 2. The robot system is constructed on Turtlebot. Turtlebot is mobile robot which can move using its own wheels. Turtlebot has a bumper to absorb bumping shock of contact. The bumper has contact sensors installed. We used Robot Operating System (ROS) to control the robot system. The robot is controlled from a notebook computer via a USB interface. Microsoft Kinect, Sound Watcher [15] and a IR camera are also connected to the notebook. Sound Watcher is a sound recognition device that recognizes sounds related to user's life, such as alarm clocks, fire alarms, and so on.

The robot works according to the behavior model illustrated in Fig. 3. The behavior model is constructed by a state transition diagram. The behavior model is constructed based on Kudo's work [16]. As illustrated in Fig. 3, the model consists of four states "Wait", "Navigation", "Human tracker", and "Touch behavior". First, in "Wait", the robot keeps to waiting at the base station until the specific sound related to user's life is recognized. Second, once recognizing the specific sound, the state transits from "Wait" to "Navigation". In "Navigation", first of all the robot looks around to detect the user. Unless detecting the user around the robot, the robot estimates the location where the user stays in a room and sets the estimated location to a destination. The robot leaves for the destination while exploring the room. Third, if the robot detects the user while in "Navigation", the state transits to "Human tracker". In "Human tracker", the robot attempts to approach the user while tracking him. If

Fig. 1 Robot system
configuration

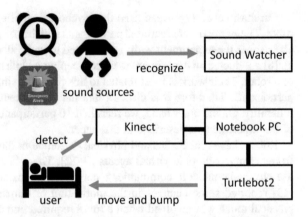

Fig. 2 Robot appearance

the robot misses the user, the state returns to "Navigation" again. In "Navigation", the robot looks around again to detect the user. Unless detecting the user the robot re-estimates the user location to reset the destination. Fourth, if the robot succeeds to approach the user in "Human tracker", the state transits from "Human tracker" to "Touch behavior". In "Touch behavior", the robot begins to bump the user. If the bumping action of the robot succeeds to notify the specific sound, the state transits from "Touch behavior" to "Navigation". In "Navigation", if there are no more sounds

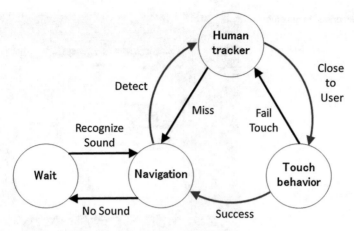

Fig. 3 State transition diagram for behavior of the hearing-dog robot

Fig. 4 Map of experimental environment (the figures in this figure denote length(cm)

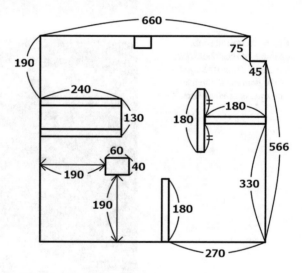

to be notified of the user, the robot sets the destination to the base station. If the robot fails to notify in "Touch behavior", the state transits back again to "Human tracker".

We conducted a simple experiment to evaluate the proposed behavior model. We had actually constructed an experimental environment in the room as shown in Fig. 4. Figure 5 is the map generated by using SLAM. In the experiment the robot could act according to the behavior model. Figure 6 shows an example of the route the robot moved.

Fig. 5 Generated map using SLAM

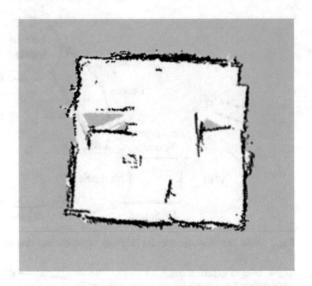

Fig. 6 Acquired route example of the robot from the base station (the *square* of *left bottom*) to the user (the *square* of *right bottom*)

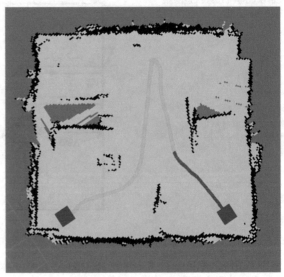

4   Research Question

In "Touch behavior", the robot begins to bump the user as mentioned in the previous section. Here we consider to design the bumping behavior effective to wake people up. Thus, this paper focused on the following research question.

RQ : What kind of robot-bumping behavior is effective to wake a sleeping person up?

From the point of view of robot control, the bumping behavior should be designed for simplicity. Turtlebot has no arms or legs. Therefore we focused on the robot-bumping force and period in order to design the robot's bumping behavior. The robot can control its own wheels and change the force and the period of bumping.

5 Experiment

5.1 Method

In the experiment a participant was sleeping on a bed, as illustrated in Fig. 7, when the robot directly approached the bed and bumped into one of its legs. The robot repeated this behavior for a certain bumping period T and force F until the participant woke up. We calculated the time from when the robot began to bump to when the participant woke up. Each participant experienced 2×2 conditions for the period T and force F on each different day in random order; $T = 1s$ or $4s$, while $F = 12N$ or $26N$. Before the experiment we confirmed that the participant was able to recognize the difference among the conditions while lying on the bed. Addition to this, 12N was minimum force which the participant could perceive while lying.

The robot began to bump the bed 40 min later after the participant went to sleep. The experiment used a sensor (Nemuri Scan NN-1100; Paramount Bed Co., Ltd.) [17] to judge whether the participant was actually asleep or not.

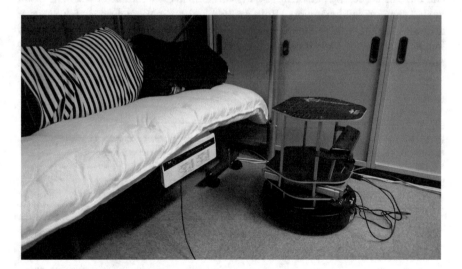

Fig. 7 Experimental environment

We recruited 16 participants (14 men, two woman, 20–25 years old) without hearing impairments. Each participant wore earplugs to muffle the sounds of the robot working and the neighboring environment. The earplugs completely muffled sounds more than 30 dB.

5.2   Results and Discussion

The robot succeeded in waking 16/16 participants up under all 2 × 2 conditions. Sekiya et al. [8] conducted the same experiment and also reported all of the participants woke up in their experiment. Thus the robot seems to have an ability to wake people up.

The average time until waking the participants up is illustrated in Fig. 8. Before the experiment, we considered that frequent and strong bumping would be effective. According to this hypothesis we believed the 1s-26N condition would give the shortest time to wake up, but the result was that the 4s-26N condition gave the shortest time. Regarding the shortest average time, Sekiya et al. also reported the same result. Addition to four conditions, two bar-graphs on the right side in Fig. 8 show the average time of a wearable device and a smartphone. The results on the wearable device and the smartphone were investigated by Sekiya et al.

Statistical test was conducted for the average time. We executed Brunner-Munzel's test between 4s-26N and the other conditions. A significant effect was found only between 4s-26N and the wearable device. The result indicates that the robot bumping can wake up a sleeping person up slightly earlier than the wearable device. The result is also same to Sekiya's work. Both 1s-26N and 4s-26N conditions show shorter average time for waking up. Meanwhile both 1s-12N and 4s-12N conditions show longer average time for waking up. The result may indicate strong condition is more effective than frequent condition. This experiment could recruited 16 participants

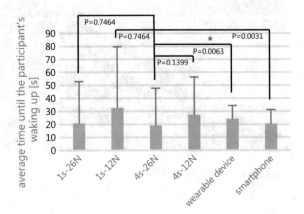

Fig. 8 Average time until the participants woke up. Significant values obtained after applying a Bonferroni correction for multiple tests

more than ten participants of Sekiya's work. But 16 participants are not sufficient to conclude the efficiency of the robot. Thus we need to recruit more participants and continue to conduct the experiment.

6 Conclusion

We presented the concept of the hearing-dog robot. Moreover we reported the experimental result to evaluate the usefulness and effectiveness of the robot. The experiment recruited 16 participants and compared the result with Sekiya's work [8]. Our result also supported Sekiya's work. On the other hand, we need much more participants to conclude the efficiency of the robot bumping.

We conducted the experiment on participants who simulated people with hearing difficulties. For our future works, we are planning to recruit deaf participants to evaluate the robot.

This work was supported by JSPS KAKENHI Grant Number 15H02768 and The Telecommunications Advancement Foundation Research Grant.

References

1. Hart, L.A., et al.: The socializing role of hearing dogs. Appl. Anim. Behav. Sci. **47**(1), 7–15 (1996)
2. Guest, C. M., et al.: Hearing dogs: A longitudinal study of social and psychological effects on deaf and hard-of-hearing recipients. J. Deaf Stud. Deaf Educ, 11(2), 252–261 (2006)
3. Matamoros, R., Seitz, L.L.: Effects of assistance dogs on persons with mobility or hearing impairments: A pilot study. J. Rehabil. Res. Dev. **45**(4), 489–503 (2008)
4. Gácsi, M., et al.: Assistance dogs provide a useful behavioral model to enrich communicative skills of assistance robots. Front. Psychol, 4 (2013)
5. Lakatos, G., et al.: Dog-inspired social behavior in robots with different embodiments. (2013)
6. Koay, K. L., et al.: Hey! There is someone at your door. A hearing robot using visual communication signals of hearing dogs to communicate intent. In: 2013 IEEE Symposium on Artificial Life (ALIFE). IEEE (2013)
7. Furuhashi, M., et al.: Haptic Communication Robot for Urgent Notification of Hearing-Impaired People. In: ACM/IEEE International Conference on Human-Robot Interaction (HRI) (2016)
8. Sekiya, D., et al.: Can a robot wake a sleeping person up by giving him or her a nudge?. In: ACM/IEEE International Conference on Human-Robot Interaction (HRI) (2017)
9. Kidd, C. D., Breazeal, C.: Effect of a robot on user perceptions. In: IEEE/RSJ International Conference on Intelligent Robots and Systems, pp. 3559–3564 (2004)
10. Kiesler, S., et al.: Anthropomorphic interactions with a robot and robot-like agent. Soc. Cogn. **26**(2), 169–183 (2008)
11. Li, J., Chignell, M.: Communication of emotion in social robots through simple head and arm movements. Int. J. Soc. Robot. **3**(2), 125–142 (2011)
12. Li, J.: The nature of the bots: How people respond to robots virtual agents and humans as multimodal stimuli. In: Proceedings of ICMI 2013, pp. 337–340 (2013)
13. Nakagawa, K., et al.: Effect of robot's active touch on people's motivation. In: ACM/IEEE International Conference on Human-Robot Interaction (HRI), pp. 465–472 (2011)

14. Chen, T. L., et al.: Touched by a robot: An investigation of subjective responses to robot-initiated touch. In: ACM/IEEE International Conference on Human-Robot Interaction (HRI), pp. 57–464 (2011)
15. Tsuzuki, H., et al.: An approach for sound source localization by complex-valued neural network. In: IEICE Transactions on Information and System E96-D(10), pp. 2257–2265 (2013)
16. Kudo, H., et al.: Behavior model for hearing-dog robot. In: International Conference on Soft Computing and Intelligent Systems and International Symposium on Advanced Intelligent Systems (SCIS & ISIS 2016) (2016)
17. Kogure, T., et al.: Automatic sleep/wake scoring from body motion in bed: Validation of a newly developed sensor placed under a mattress. J. Physiol. Anthropol. **30**(3), 103–109 (2011)

Implementation of Document Production Support System with Obsession Mechanism

Ziran Fan and Takayuki Fujimoto

Abstract Under the development of the IT introduction in enterprises, the efficiency of works is required more than ever. It means that not only the enterprises' measures that apply the IT into the business operations, but also the worker's ability to carry out the works with IT effectively is essential. This paper considers the work efficiency in people's work style today, and focuses on the task of document producing. The efficiency of document producing ithe essential element in all the business activities, and it is the task that everyone can achieve simply. Enhancing the efficiency of document producing is related to the improvement of the work efficiency in users' work style. In late years, producing documents at mobile system is becoming more popular by the appearance of tablet computers and large-screen smartphones, the efficiency of document producing is even more required. At this paper, we focus on the most essential point in producing document, which is "to complete within a deadline". To implement that function into the system, we apply a human cognition mechanism for deadline, such as the mechanism of obsession to deadline. We call it as "Deadline Obsession (DLO)" at this paper and use that cognition mechanism to implement a support system for document producing to improve users' work efficiency.

Keywords IT · Information design · Work efficiency · Document production · Cognition mechanism · Deadline obsession · Media system

1 Introduction

Today, people's lifestyle is completely changed by the evolution of information technology. IT is applied throughout people's daily lives and makes them even more convenient. Needless to say, diffusion and application of IT do not only affect people's

Z. Fan (✉) · T. Fujimoto (✉)
Graduate School of Information Sciences and Arts, Toyo University, Kawagoe-shi, Japan
e-mail: swterc@gmail.com

T. Fujimoto
e-mail: fujimoto@toyo.jp

© Springer International Publishing AG 2018
R. Lee (ed.), *Computational Science/Intelligence and Applied Informatics*,
Studies in Computational Intelligence 726, DOI 10.1007/978-3-319-63618-4_5

lifestyles but also are influential in the business fields. The evolution of information technology in enterprises makes the difference between the "IT company" and others obscure lately. Also, even the fishing industry and the farming industry that are known as traditional business are applying IT to their trades. IT became more and more important to the business regardless of fields and categories.

"Efficient business operations" is the greatest subject of enterprises' system of IT and could be considered by 2 aspects. One is "Efficiency of the whole business operations in enterprise". It refers to the compression of the unnecessary business and the curtailment of the execution time at necessary works. These require the scheme involved with enterprises' management structure such as the arrangement of business flow and the improvement of implement's performance. On the other hand, the second aspect is "Efficient works in workers' work styles". The system to Introduce IT into the business operations is not perfect for the introduction of IT. The skills and abilities of workers to execute works effectively by IT are also required. In short, under the background of the IT' s introduction in enterprises, how workers apply IT effectively in their work style is a key to business success. It is impossible to attempt the whole business efficiency in enterprise without enhancing the efficiency of works that individual worker is carrying out.

This paper aims at the support of the work efficiency in workers' work styles. Therefore, we ought to specify the most important task in workers' work styles and also the most required task to enterprises. The support does not require the skills with high degree of expertise or the new abilities that workers have to acquire. It refers to a measure to enhance the work efficiency with the capacity that workers already have.

We focus on the document producing that applies computer technology to handle the documents. It can be said that the task of document producing is the typical example of the deskwork in enterprise. Document producing such as creating presentation documents, schedule management, regulating the references, is an essential task to enterprises in all business fields and places of work. The business such as sales of goods or product's development may be considered as the work operations that are unrelated to the document producing. However, their execution processes all involve the task of document producing. The effect of Document producing for enterprises is significant and its relation with business activities is everywhere. Everyone can carry out the task of document producing easily with the general ability, nevertheless all posts and positions require the skill of document producing in enterprise. The work efficiency of document producing is closely associated with workers' work efficiency.

This paper proposes and develops a support system to enhance the efficiency on document producing. We take up elements of the efficiency on document producing and analyze them. Especially, we use a human's general cognition mechanism for the deadline. We call this cognition mechanism as "Deadline Obsession" (DLO). We would explain the idea of DLO and also devises and creates a support system of document production using DLO mechanism [3].

2 Background

According to the "Survey for IT Innovator" in 2016 by NikkeiBP, on the 157 targets, 70% of targets shows the active attitude toward the introduction and system of IT through the increase of the digital investment [1]. And the rate of the targets that increased the investment over 20 is 24.8%. This data shows the close relationship between IT and enterprises management behind the active attitude to invest. For today's enterprises, IT does not have the big additional value itself. How to apply IT to make the difference and the superiority on the competition is a serious subject.

Absolutely, for enterprises, the introduction of IT is not the purpose itself. IT is applied for some purpose or reason that enterprise has. Then, why so many enterprises have concern in the introduction of IT? According to the "Enterprise IT Trend Survey 2016", the subject of "efficient business operations" is the greatest purpose regarding the IT investment in enterprises [2].

It is obvious that enterprises desire to enhance the business efficiency by the introduction of IT. However, the result is not satisfactory. According to the "A questionnaire about the work efficiency" by EnJapan, 66% of answers (workers) thought they could probably carry out works more effectively. Moreover, 30% of answers related the reason of inefficient works is the abilities and skills of themselves of works. Moreover, 30% of answers indicates that the reason of inefficient works is their abilities and skills for works.

There is a big gap between the enterprises, which devote a great deal of attention to introduce IT for business efficiency and workers, who are distressed for the problem that they hardly reflect the effect of IT for the efficiency of their actual works. It would be difficult to realize the efficient business operations in enterprises without the solution of this gap.

3 Purpose

3.1 The Work Efficiency in Workers' Work Styles

There are two main meanings of the efficient business operations in enterprises. It refers to the compression of the unnecessary business and the curtailment of the execution time at necessary works. Enterprises commonly aim at the curtailment of the execution time, which is necessarily expected with IT introduction.

However, there are some problems on this measure. At first, one of the problems is the application of IT to the business operations. Enterprises may have different business operations by the different targets (consumer, company or government) even in the same business field, and the type of IT system would be changed depending on those differences. For enterprises, how to apply IT effectively in business operations for the realization of the work efficiency is a hard subject.

Moreover, the equipment of IT environment is also a problem. The standard of preparation for the suitable IT environment for each enterprise's business operations. Also sometimes it requires immense investment of enterprises.

The last problem is the most serious. It is about "human". Even though the IT environment is consummated and the system of the business operations is perfect, if workers cannot apply those to the works effectively, all of them would be useless. This is certainly the disparity between "the efficiency in the introduction of IT in enterprises" and "the efficiency in workers' work styles by IT" which is described earlier. Enhancing the work efficiency in workers' work styles would be related to the efficiency of the whole business operations in enterprises. Conversely, the efficiency of the work operations in enterprises would be impossible without the realization of the work efficiency in workers' work styles.

It is necessary to consider the efficiency from workers' point of view, but the view of enterprises. There are three aspects to be considered.

At first, we need to consider the work that IT should be applied for. It must be the basic task that workers are generally carrying out, and also should have a necessity for enterprise's business.

Second, the object of IT that workers apply to work is required. It should be a tool such as a device or software, and has great usability and practical effect for the workers' works.

Finally, the aspect to be considered is the workers' ability. It refers to the work applicable by universal knowledge or skills that everyone has and should not require high-level professional knowledge of workers.

The method that meets those requirements would be a key to solve the disparity of the work efficiency between the enterprises and workers. After consideration, this paper focuses on document producing and proposes a support system to enhance the efficiency of users' works.

3.2 The Relationship Between Document Producing and Work Efficiency

Making good use of information is the most preferential object for enterprises' system of IT. Handling the documents with computers is the work operation most nearly concerned with information in enterprises. Absolutely, information contains sound, image and video. Characters mainly used in document producing is only one kind of it. However, according to the examination, characters is the information applied in enterprises the most commonly and also it is used in workers' work styles the most widely.

In the business activities such as filling documents, collecting materials, and planning new enterprises, the work operation mainly targeting character information for the document producing is essential. For example, planning enterprises is considered as the work operation, which is unrelated to the document producing. However, the required works of business processes at planning enterprises, like marketing research, information-gathering, devising ideas and creating reports, all involve document producing. In other words, the business activities are impossible without document producing. The work efficiency of document producing is a big factor that relates to the workers' work efficiency.

Document producing is an effective task because it interacts with most work operations in enterprises. It is also a simple task that can be carried out just with a computer. The feasibility to develop the handy tool of document producing which does not require a complex equipment of IT environment and can be readily available is one of the reasons we decided to propose a support system of document producing at this paper.

The workers do not need a high-level ability, the standard skill is enough for document producing. The concept of the support system of document producing we propose in this paper focuses the ability that workers already have instead of requiring the new skill to workers. It utilizes that ability effectively to enhance workers' work efficiency by supporting the work of document producing on the system.

3.3   The Realization of the Work Efficiency

The work efficiency refers to the curtailment of the execution time of works. To implement the curtailment of the execution time at document producing, it is the practical way that users complete works more quickly to shorten the execution time directly.

To shorten the execution time, users' ability and the grasp of the work are required. The document production itself is not a difficult task to solve. On the other hand, even the users who get accustomed to the document producing and have enough ability of the document production cannot always work effectively. The problem is not only users' ability but also is the way that they show the ability according to the purpose, and there is very much a possibility that users show it in the wrong way.

The biggest cause that makes the execution time prolonged of document producing is users' consciousness to works. Non-efficiency of works could be happened by the erroneous consciousness of the users such as "That can wait until tomorrow" or they can be distracted by other things to spoil their concentration on works. Keeping the users more conscious of works would lead to the curtailment of the execution time and work efficiency. It refers to the way to evoke users' consciousness by some sort of extrinsic stimuli is required.

This paper considers that users' consciousness of works is the biggest factor that affects the work efficiency in document producing. For preparation, we ought to examine how the factors affect users' consciousness of works.

3.4 Two Factors that Affect Users' Consciousness

We pick up the factors that affect users' consciousness when they are working on document production, and analyze them.

At first, external factors are considered. For example, the incentive such as the bonus or the promotion can be given when works are completed, also the penalty such as a salary cut or reproof may be given when works are failed. However, the impact of those effects to human is finite. When a human experiences the incentive or penalty continuously, the praxis will be arisen and the effect will be fallen.

Therefore, employing the internal factors inside of the users instead of the external factors is more efficacious for the work efficiency. For one thing, the internal factor of the users could be cited as the responsibility such as the feeling to complete the work and make it successful. However, Giving the users a responsibility or evoking it is not facile. It is difficult to apply to the support system we propose.

For that reason, we focus on another internal factor, that is the pressure from works. This factor appears with or without the incentive or the penalty. It refers to an obsession that everyone would experience when works are given and the deadline and expectation results are set. More or less, if the users could not complete works, they would self-hate or be ashamed. This study develops the system for work efficiency with that obsession.

3.5 What Does "DLO" Mechanism Mean?

Obsession is the universal cognitive phenomenon of human. Generally, obsession is defined as follows.

- It refers to a condition that human can recognize the consciousness as irrationality and meaningless, but they cannot remove and it springs to mind without the personal decision. (Reference of Encyclopædia Britannica).

It is different from the incentive or penalty. The praxis would not be happened and the effect would not be fallen because it is not affected by external elements.

The feature of obsession is that human cannot remove the image when it springs to mind even though they decide not to think about it. Being stick to the place and form of the object (whether it is symmetric or not, feeling like the place is changed from yesterday, and etc.) is the representative example. Obsession is the cognitive phenomenon that everyone has, so it could be effective in spite of the sexuality, age, or cultural settings of the users [9, 10].

Examining the obsession in works of document producing, this paper puts the focus on the obsession that everyone would have universally for the deadline set on works or tasks. We call it as Deadline Obsession (DLO) in this paper.

When the deadline is set for works, the users would fell pressure by the time limit (they must complete the work before that) of works, and be more conscious of the progress of works. Conversely, if the users cannot meet the deadline, it may inflict a loss to the organization. Moreover, a sense of guilt: "I could not meet the deadline" will be put into users' mind.

In this way, the obsession (DLO) is the universal cognitive phenomenon of human and we can expect a significant effect that users would perform works initiatively to complete tasks within a period. This paper uses the mechanism of DLO to design the system and strives for users' work efficiency in document producing [8, 11].

4   Implementation of System

4.1   Overview of System

This paper implements a support system of document producing to enhance the efficiency of document producing using DLO mechanism. We used integration development software—Xcode with Objective-C and implemented the system for Apple iPad.

Tablets such as iPad is recently becoming popular. It is seen in various areas that people use tablets instead of conventional computers for work, because tablets are convenient to carry and have high-performance same as computers. The sphere of users' work activity could be spread and the flexible everywhere workplace-styles could be implemented by use of tablets.

We put focus on the deadline of the document producing in systems design. Noticing the deadlines and managing the schedule at the same time is the main trend of work styles today. However, people may not remember the deadlines properly enough and they may not manage the tasks for the efficient work while suffering from DLO.

Even if they remember the deadlines on time-line, the unbalanced progress and the variation of completion will be occurred and they cannot balance multiple tasks according to the deadlines efficiently [4].

Therefore, this system enables the users to become aware of the deadlines visually while engaging in the works of document producing, and provides the work environment which links works with deadlines. We propose the system which results efficiency improvement by allowing the users to be aware of the deadlines and adjust the tasks to meet the deadlines for better progress.

The principle of the system is the development of support system of document producing; mainly treat the character information with the function of producing and editing documents. Moreover, we put the focus on the relationship between the deadlines and the process of works, and use DLO mechanism to evoke the consciousness of users [5, 6].

To implement that, we incorporate the factors: informing the deadline to users, the suitable user interface design, and the motivational trigger to "make users meet the deadlines", or "get users think he or she has to complete works" based on DLO mechanism into the system.

4.2 Functions of System

Figure 1 shows the function of loading a file or creating a new file.

When the user creates a new file, the system requires the user to set the deadline before starting the work (Fig. 2). The time before the deadline will be automatically

Fig. 1 Start system

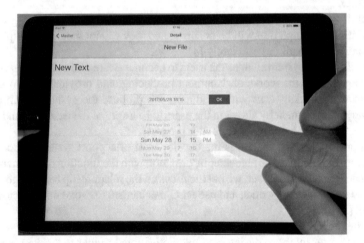

Fig. 2 Set the deadline of task

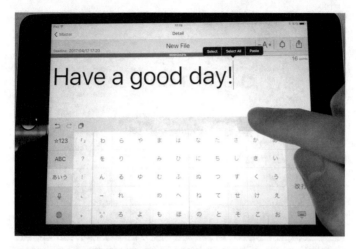

Fig. 3 Document processing

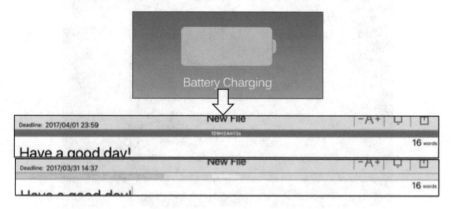

Fig. 4 Design of the deadline reminder

calculated after the settings for the date of the deadline by the user. The data of the time is stored in the file that the user created, and it will be automatically loaded when the system is started.

The work screen will be displayed when the file is chosen (Fig. 3). The work of the document producing will be mainly performed on this screen. This system stylizes a simple and convenient design for document producing specifically.

The remarkable feature of this system is a method to express the deadlines. It does not just allow users to realize the deadlines intuitively, but also it enables the users to understand how much time is left until the deadline. This is fundamental to invoke users' realization about the deadlines with degrees of the works' completion [7]. This paper uses a design of the smartphone battery as a motif for the reminder to express the remaining time before the deadline (Fig. 4).

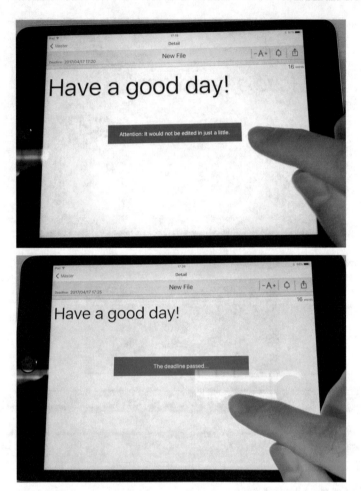

Fig. 5 Information notice of the deadline

The closer it gets to the deadline, the lower the remaining battery level becomes. Then the system informs the users of the remaining time depending on the lapse of time before the deadline (Fig. 5).

However, this cannot enable users to pay attention to the deadline and to manage the tasks according to the deadlines completely. We implemented a mechanism to function DLO in the system to allow the users to realize the deadlines to be kept more intensively. There tends to be a time lag between the deadlines set by the users and the actual deadlines to submit the documents.

Fig. 6 Motivating with DLO

Fig. 7 Changing font size

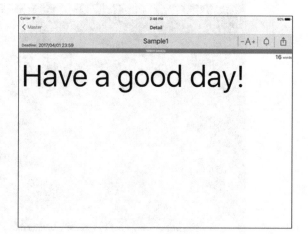

Thereby, a halfhearted idea will be occurred because of a thought: "even if the deadline passes I can still finish the work before the actual deadline, so it is ok." The consciousness with pressure such as "I must complete the work before the deadline" turns to the slackness for the consciousness just like "It is fine to finish the work at the last minute". This kind of idea is the main factor to make the situations in which people cannot keep the deadlines and it damages the functionality of DLO.

To prevent that from happening, a stronger obsessional idea is required and we assume that it is a pressure in this paper. It is essential to motivate the users to finish the work definitely before the deadline. We incorporated a mechanism to stop the users from continuing the work if the deadline passes, into the system. It is a function to unable the users to edit the file after the deadline (Fig. 6).

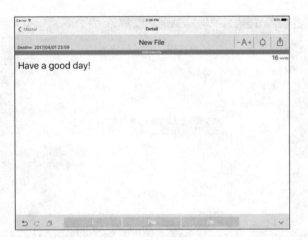

Fig. 8 Word count

Fig. 9 Interlocking with "Reminder" apps

This function brings pressure to the users and allows them to advance the works in planned ways with the realization of the deadlines. Enabling users to feel pressure would make users perform works initiatively and it results in the curtailment of the execution time and can enhance the overall efficiency of the tasks.

To improve the user experience of works of document producing by using this system, we append the additional function such as changing characters' size (Fig. 7), counting the number of characters (Fig. 8). For enhancing the applicability to works, an interlocking function with [Reminder] which is the primary system of iPad (Fig. 9), and a function that offers users can export the documents as PDF file would be also appended into the system.

Specially, users can set the reminder for the deadlines of the work on which users are working on by the interlocking function, and the system also allows users to recognize the deadline and check the progress of the work by the alert function of the reminder when the deadline passes.

5 Conclusion

5.1 Review

Today, introduction of IT for enterprises is remarkable among the various industry types, categories, and fields. It is also can be said as the popularization of the introduction of IT in workers' work style. "Efficient business operations" is the greatest purpose of enterprises' system of IT.

However, it is not enough for efficient business operations just to apply IT to enterprises' business. The abilities of IT that users utilize it to achieve the efficient results of work are also essential. To implement "the efficient business operations in enterprises by IT", it requires the solution about "the efficient works in workers' work styles by IT system".

This paper examined three aspects from workers' work styles; the specification of work which workers apply IT to, the feasibility of the tool workers utilize, and making most of workers' ability on works. After consideration, this paper focused on document production handled by computer.

Document producing is indispensable to any enterprise, and it treats the characters that is information most commonly used in enterprises and it can be applied to any work. Moreover, document producing is the essential task in workers' work styles, and has a significant effect to workers' work efficiency.

The tool used for document producing is simple, and document producing does not require workers with the special skills, so workers could carry out that task simply with the ability what they already have. Based on consideration, this paper proposed the support system of document producing as a tool that be serviceable for the work efficiency in users' work styles and implemented that.

The most effective way to implement the work efficiency on document producing is evoking users' consciousness of works. We analyzed the external and internal factors that affect users' consciousness. We put the focus on the pressure of works, which is the obsession, the universal cognitive phenomenon of human, and proposed the DLO (Deadline Obsession) that is the most influential obsession on document producing for work efficiency of users.

In the system, we used a design of the smartphone battery as a motif to express the deadline to allow users to realize the deadlines intuitively, and also incorporated the DLO mechanism, which allows users to have the strong motivation to complete works before the deadlines, into the system.

5.2   *Future Works*

We describe about the future issues of this paper.

First, the accurate evaluation and verification of DLO mechanism are required. DLO mechanism is a universal cognitive phenomenon of human, but it does not always affect all users at the same degree. Therefore, it is necessary to examine various kinds of users with different individualities. For that, the verification of system's utilization effect and a thorough audit of DLO mechanism by the long-term observation of users' actual usage for works needs to be performed as necessary. Based on their results, we will consider the most suitable way to use the system of DLO mechanism and how to implement the system.

Second, we will improve the system from the results of the survey. To realize the continuous use and the applicability to the specific works, the addition of necessary functions is required.

Finally, we plan to release the system as a product. The purpose of the system is to be a useful tool for the work efficiency in users' work styles. Moreover, the system can be used internationally by the release. Making most of this opportunity, we will examine the current situation of different work styles of people by the different regions, cultures and economic circumstances and explore the internationally common ideal work style. That leads to the development of the future works of this study.

Acknowledgements   This work was supported by JSPS KAKENHI Grant Number 17K00730.

References

1. Nikkei Computer: Survey of IT innovator 2016 (2016)
2. EnJapan: A questionnaire about the work efficiency (2015)
3. Fan, Z., Fujimoto, T.: A Text Editor Apps for Work Efficiency used by "DLO: Deadline Obsession". In: The 185th SIG–ICS (2016)
4. Fujimoto, T.: Design that makes information easy: What is information design and what is not information design. Inf. Sci. Technol. **65**(11), 450–456 (2015)
5. Moribe, Y.: A Study Note: Information Design and Its Directionality, vol. 12, no. 1, pp. 289–300. Bulletin of Miyazaki Municipal University Faculty of Humanities (2005)
6. Kikuchi, T., Itoh, T., Okazaki, A.: Recent trend of information design and information visualization studying from web navigation techniques. Soc. Art Sci. **4**(1), 1–12 (2005)
7. Sasa, M.: Direction of information design in using services. Rev. Artic. (2011)
8. Kubota, Y., Kokaji, S.: The trouble of smartphone network that makes a mess of life (2014)
9. Hospital Psychopathology, Behavior Therapy Lab, What is the Obsessive-compulsive Disorder (OCN) (2015)
10. Kurabayashi, S.: The concept of health and considerations on its modern connotations–from a viewpoint on the theory of happiness, vol. 4, pp. 1–10. Bulletin of Takasaki University of Health and Welfare (2005)
11. Syudo, Y., Tanno, Y.: Relationship between obsession and chimerical idea of healthy people: Analyzing with Path Analysis, Jpn. Soc. Pers. Psychol. **11**, 66–67 (2002)

Detecting Outliers in Terms of Errors in Embedded Software Development Projects Using Imbalanced Data Classification

Kazunori Iwata, Toyoshiro Nakashima, Yoshiyuki Anan and Naohiro Ishii

Abstract This study examines the effect of undersampling on the detection of outliers in terms of the number of errors in embedded software development projects. Our study aims at estimating the number of errors and the amount of effort in projects. As outliers can adversely affect this estimation, they are excluded from many estimation models. However, such outliers can be identified in practice once the projects have been completed; therefore, they should not be excluded while constructing models and estimating errors or effort. We have also attempted to detect outliers. However, the accuracy of the classifications was not acceptable because of a small number of outliers. This problem is referred to as data imbalance. To avoid this problem, we explore rebalancing methods using k-means cluster-based undersampling. This method aims at improving the proportion of outliers that are correctly identified while maintaining the other classification performance metrics high. Evaluation experiments were performed, and the results show that the proposed methods can improve the accuracy of detecting outliers; however, they also classify too many samples as outliers.

K. Iwata (✉)
Department of Business Administration, Aichi University, 4-60-6, Hiraike-cho, Nakamura-ku, Nagoya, Aichi 453-8777, Japan
e-mail: kazunori@vega.aichi-u.ac.jp

T. Nakashima
Department of Culture-Information Studies, Sugiyama Jogakuen University, 17-3 Moto-machi, Hoshigaoka, Chikusa-ku, Nagoya, Aichi 464-8662, Japan
e-mail: nakasima@sugiyama-u.ac.jp

T. Nakashima
Institute of Managerial Research, Aichi University, 4-60-6, Hiraike-cho, Nakamura-ku, Nagoya, Aichi 453-8777, Japan

Y. Anan
Base Division, Omron Software Co., Ltd., Higashiiru, Shiokoji-Horikawa, Shimogyo-ku, Kyoto 600-8234, Japan
e-mail: yoshiyuki_anan@oss-g.omron.co.jp

N. Ishii
Department of Information Science, Aichi Institute of Technology, 1247 Yachigusa, Yakusa-cho, Toyota, Aichi 470-0392, Japan
e-mail: ishii@aitech.ac.jp

© Springer International Publishing AG 2018   65
R. Lee (ed.), *Computational Science/Intelligence and Applied Informatics*,
Studies in Computational Intelligence 726, DOI 10.1007/978-3-319-63618-4_6

Keywords Embedded software · Imbalanced dataset · Support vector machine · k-means clustering algorithms · Undersampling

1 Introduction

The growth and expansion of our information-based society has resulted in an increasing number of information products. In addition, the functionality of these products is becoming ever more complex [6, 14]. Guaranteeing the quality of software is particularly important because it relates to reliability. Therefore, it is increasingly important for corporations that develop embedded software to implement efficient processes while guaranteeing timely delivery, high quality, and low development costs [2, 12, 15, 16, 18–21]. Companies and divisions involved in developing of such software focus on a variety of improvements, particularly in their processes. Estimating the number of errors and the amount of effort is necessary for new software projects and guaranteeing product quality is particularly important because the number of errors is directly related to the product quality and the amount of effort is directly related to cost, which affect the reputation of the corporation. Previously, we investigated the estimation of total errors and effort using an artificial neural network (ANN) and showed that ANN models are superior to regression analysis models for estimating errors and effort in new projects [8, 9]. We proposed a method to estimate intervals for the amount of effort using a support vector machine (SVM) and an ANN [7, 10]. These models were constructed with data that excluded outliers. The outliers can be identified in practice once the projects have been completed. Hence, they should not be excluded while constructing models and estimating effort. We attempted to classify embedded software development projects based on verifying whether the amount of efforts was an outlier using an ANN and SVM [11]. However, the accuracy of the classifications was not acceptable because of a small number of outliers. This problem occurs in most machine learning methods and is referred to as data imbalance. It exists in a broad range of experimental data [1, 22]. Data imbalance occurs when one of the classes in a dataset has a very small number of samples compared to the number of samples in other classes. When the number of instances of the majority class exceeds that of the minority class by a significant amount, most samples are classified into a class to which the majority samples belong. Therefore, the number of the outliers is small, and they are classified as normal values. To avoid this problem, we explored rebalancing methods in terms of errors using k-means [5] cluster-based undersampling. Evaluation experiments were performed to compare the classification accuracy using k-means undersampling with that of random undersampling and no undersampling using ten-fold cross-validation.

2 Related Work

2.1 Undersampling

Undersampling is one of the most common and straightforward strategies for handling imbalanced datasets. Samples of the majority class are dropped to obtain a balanced dataset. Simple undersampling randomly drops samples to generate a balanced dataset.

2.2 Cost-Sensitive Learning

Unlike cost-insensitive learning, cost-sensitive learning is a type of learning that considers misclassification costs [17]. Additionally, cost-sensitive learning imposes different penalties for different misclassification errors. It aims at classifying samples into a set of known classes with high accuracy. Cost-sensitive learning is a common approach that solves the problem associated with imbalanced datasets.

2.2.1 Cost-Sensitive SVMs

SVMs have proven to be effective in many practical applications. However, the application of SVMs has limitations when applied to the problem of learning from imbalanced datasets. A cost-sensitive SVM, which assigns different misclassification costs, is good solution to address the problem [3, 13]. Such an SVM is develped using different error costs for the positive and negative classes, and can improve classification accuracy for a small number of classes.

2.3 Our Contribution

The above algorithm has a certain level of classification accuracy for some imbalanced datasets; however, it cannot improve the accuracy for highly imbalanced datasets. Therefore, in this research, we proposed a rebalancing method using k-means cluster-based undersampling.

3 Datasets and Outliers

3.1 Original Datasets

Using data from a large software company, the classification methods divide the number of anticipated errors into normal values and outliers. The data consist of the following features:

Class:
  This indicates whether the total number of errors for an entire project is a normal value or an outlier. Predicting this value is the objective of the classification.
Volume of newly added steps (V_{new}):
  This feature denotes the number of steps in the newly generated functions of the target project.
Volume of modification (V_{modify}):
  This feature denotes the number of steps modified or added to existing functions that were needed to use the target project.
Volume of the original project (V_{survey}):
  This feature denotes the original number of steps in the modified functions and the number of steps deleted from the functions.
Volume of reuse (V_{reuse}):
  This feature denotes the number of steps in a function of which an external specification is only confirmed and which are applied to the target project design without confirming the internal content.

3.2 Determination of Outliers

This study examined the classification of outliers in terms of the number of errors in a project. Fig. 1 shows the distribution of the number of errors, whereas Fig. 2 is a boxplot of this metric. The lowest datum of the boxplot is 0, which is the lowest possible number of errors in the projects and higher than 1.5 times the interquartile range (IQR) of the lower quartile. The highest valid datum is within 1.5 times the IQR of the upper quartile. The outliers are denoted by circles. Here, the values are spread along the Y-axis to more clearly present the distribution of the outliers; however, the Y-coordinate has no other meaning. Of the total of 1,419 data points, 143 are outliers. Detailed values of the boxplot are listed in Table 1.

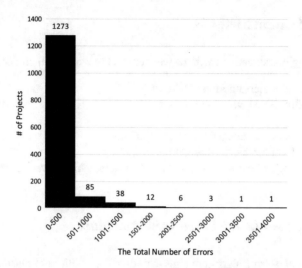

Fig. 1 Distribution of the total number of errors (in intervals of 500 errors)

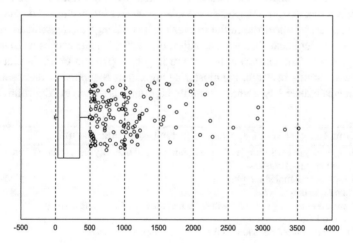

Fig. 2 Boxplot of the number of errors

Table 1 Detailed information of the boxplot shown in Fig. 2

	IQR	Minimum	Lower quartile	Median	Upper quartile	Maximum
Values	189.0	0.0	35.0	86.0	224.0	507.5

4 Classification Methods

The following classification methods were created to compare their accuracy:

- SVM without undersampling (SVM w/o).
- Cost-sensitive SVM without undersampling (CSSVM).
- SVM with random undersampling (W/Random).
- SVM with k-means cluster-based undersampling (W/n Clusters), where n is the number of clusters, which is varied from 2 to 15.

4.1 K-Means Cluster-Based Undersampling

The k-means clustering algorithm aims at finding the positions of clusters that minimize the distance from the data points to k clusters. The algorithm is often presented as a method that assigns samples to the nearest cluster by distance. The main steps of k-means are to select the initial cluster centers, change the classification of the data based on Euclidean distance and adjust the cluster centers according to the classification result. The clustering results are largely dependent on the initial cluster assignment. In this research, the clustering algorithm is applied to undersampling. The k-means cluster-based undersampling algorithm is shown in Algorithm 1.

Algorithm 1 Algorithm of K-Means Cluster-Based Undersampling

Require: L: majority class samples from a training set, S: minority class samples from a training set, k: the number of clusters.
Ensure: L_u: under-sampled samples
1: Initialize L_u to empty-set
2: Use k-means clustering to form clusters on L denoted by C_i , where $1 < i \leq k$
3: **for** $i = 1$ to k **do**
4:    Calculate the number of samples for all the clusters denoted by:

$$n_i = \frac{|C_i|}{|L|} \times |S|$$

   , where $|X|$ indicates the number of elements in X
5:    Set T to n_i elements selected randomly from C_i
6:    Assign the union L_u and T to L_u
7: **end for**
8: **return** L_u

---

5 Evaluation Experiment

5.1 Data Used in the Evaluation Experiment

To evaluate the performance of the proposed technique, we performed ten-fold cross-validation on data from 1,419 real projects. The original data were randomly partitioned into 10 equally sized subsamples (each subsample having data from 141 or 142 projects). One of the subsamples was used as the validation data for testing the model, while the remaining nine subsamples were used as training data. The cross-validation process was repeated 10 times, with each of the 10 subsamples used exactly once as validation data. An example of ten-fold cross-validation is shown in Fig. 3.

5.2 Evaluation Criteria

This study focused on the imbalance problem wherein the minority class (outliers) has much lower precision and recall than the majority class (normal values). Accuracy metrics place more weight on the majority class than on the minority class, which makes it difficult for a classifier to perform well on the minority class.

By convention, the class label of the minority class is positive, whereas that of the majority class is negative. The True Positive (*TP*) and True Negative (*TN*) values, as summarized in Table 2, denote the number of positive and negative samples

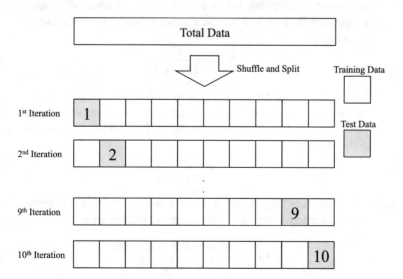

Fig. 3 Ten-fold cross-validation

Table 2 Confusion matrix

	Predict positive	Predict negative
Actual positive	TP	FN
Actual negative	FP	TN

that are correctly classified, while the False Positive (*FP*) and False Negative (*FN*) values denote the numbers of samples incorrectly classified as positive and negative, respectively.

The following eight criteria were used as performance measures for the classification methods. The best value of these criteria is 1.0, whereas the worst is 0.0.

(1) Accuracy (*ACC*) is the proportion of the total number of predictions and is calculated as the number of all correct predictions divided by the total number of samples using the following equation:

$$ACC = \frac{TP + TN}{TP + TN + FP + FN}. \tag{1}$$

(2) Precision (*PREC*) is the proportion of correctly predicted positive cases. It is calculated as the number of accurate positive predictions divided by the total number of positive predictions using the following equation:

$$PREC = \frac{TP}{TP + FP}. \tag{2}$$

(3) Sensitivity (*SN*, recall, or *TP* rate) is the proportion of positive cases that are correctly identified. It is calculated as the number of correct positive predictions divided by the total number of positives using the following equation:

$$SN = \frac{TP}{TP + FN}. \tag{3}$$

(4) Specificity (*SP* or *TN* rate) is defined as the proportion of negative cases that are correctly classified and calculated as the number of correct negative predictions divided by the total number of negatives using the following equation (4):

$$SP = \frac{TN}{TN + FP}. \tag{4}$$

(5, 6, 7) The F-measure (F_β) is the harmonic mean of precision and sensitivity. It is calculated as a weighted (β) average of the precision and sensitivity as follows:

$$F_\beta = \frac{(1 + \beta^2) \times PREC \times SN}{\beta^2 \times PREC + SN}. \tag{5}$$

Table 3 Classification results obtained using the support vector machine without undersampling (SVM w/o) Method

		Predicted classes	
		Outliers	Normal values
Actual classes	Outliers	65	78
	Normal values	16	1260

Table 4 Classification results obtained using the cost-sensitive SVM without undersampling (CSSVM) Method

		Predicted classes	
		Outliers	Normal values
Actual classes	Outliers	26	117
	Normal values	0	1276

F-measures $F_{0.5}$, F_1, and F_2 are commonly used. The larger β, the more importance sensitivity has in the equation.

(8) The G-measure (G) is based on the sensitivity of both the positive and negative classes, calculated as follows:

$$G = \sqrt{SN \times SP}. \tag{6}$$

This paper aimed to detect all of outliers; however, there is a trade-off between precision and sensitivity. In generally, precision improves at the expense of sensitivity and sensitivity improves at the expense of precision [4]. Thus, improving SN, F_2 and G was important while maintaining the other classification performance metrics high.

5.3 Results and Discussion

For each method described in Sect. 4, the confusion matrices of the experimental results for all projects using ten-fold cross-validation are presented in Tables 3, 4, 5, 6, 7, 8, 9, 10, 11, 12, 13, 14, 15, 16, 17, 18, 19 and 20. The values in the tables represent the aggregate over 10 experiments. The results in the tables are summarized in Figs. 4, 5, 6, and 7. High values in Figs. 4 and 5 mean high accuracy. In contrast, low values in Figs. 6 and 7 suggest high accuracy.

Table 20 summarizes the results of the criteria for all methods. The underlined values indicate the best results. The top five methods for each criterion are indicated in bold type. The results of the criteria for SVM w/o, CSSVM, W/Random and W/14 Clusters appear in Fig. 8. The results of W/n Clusters except W/14 Clusters are omitted because these show similar features.

Table 5 Classification results obtained using the SVM with random undersampling (W/Random) Method

		Predicted classes	
		Outliers	Normal values
Actual classes	Outliers	115	28
	Normal values	137	1139

Table 6 Classification results obtained using the W/2 clusters method

		Predicted classes	
		Outliers	Normal values
Actual classes	Outliers	131	12
	Normal values	302	974

Table 7 Classification results obtained using the W/3 clusters method

		Predicted classes	
		Outliers	Normal values
Actual classes	Outliers	130	13
	Normal values	309	967

Table 8 Classification results obtained using the W/4 clusters method

		Predicted classes	
		Outliers	Normal values
Actual classes	Outliers	136	7
	Normal values	301	975

Table 9 Classification results obtained using the W/5 clusters method

		Predicted classes	
		Outliers	Normal values
Actual classes	Outliers	131	12
	Normal values	310	966

Table 10 Classification results obtained using the W/6 clusters method

		Predicted classes	
		Outliers	Normal values
Actual classes	Outliers	133	10
	Normal values	299	977

Table 11 Classification results obtained using the W/7 clusters method

		Predicted classes	
		Outliers	Normal values
Actual classes	Outliers	137	6
	Normal values	313	964

Table 12 Classification results obtained using the W/8 clusters method

		Predicted classes	
		Outliers	Normal values
Actual classes	Outliers	131	12
	Normal values	312	964

Table 13 Classification results obtained using the W/9 clusters method

		Predicted classes	
		Outliers	Normal values
Actual classes	Outliers	133	10
	Normal values	299	977

Table 14 Classification results obtained using the W/10 clusters method

		Predicted classes	
		Outliers	Normal values
Actual classes	Outliers	133	10
	Normal values	311	965

Table 15 Classification results obtained using the W/11 clusters method

		Predicted classes	
		Outliers	Normal values
Actual classes	Outliers	132	11
	Normal values	292	984

Table 16 Classification results obtained using the W/12 clusters method

		Predicted classes	
		Outliers	Normal values
Actual classes	Outliers	137	6
	Normal values	306	970

Table 17 Classification results obtained using the W/13 clusters method

		Predicted classes	
		Outliers	Normal values
Actual classes	Outliers	130	13
	Normal values	299	977

Table 18 Classification results obtained using the W/14 clusters method

		Predicted classes	
		Outliers	Normal values
Actual classes	Outliers	134	9
	Normal values	289	987

Table 19 Classification results obtained using the W/15 clusters method

		Predicted classes	
		Outliers	Normal values
Actual classes	Outliers	134	9
	Normal values	321	955

Table 20 Accuracy comparison for all methods

	ACC	$PREC$	SN	SP	$F_{0.5}$	F_1	F_2	G
SVM w/o	**0.9338**	**0.8025**	0.4545	**0.9875**	**<u>0.6959</u>**	**0.5804**	0.4977	0.6700
CSSVM	**0.9175**	**1.000**	0.1818	**<u>1.0000</u>**	0.5263	0.3077	0.2174	0.4264
W/Random	**0.8837**	**0.4563**	0.8042	**0.8926**	0.4996	**<u>0.5823</u>**	**<u>0.6978</u>**	0.8473
W/2 Clusters	0.7787	0.3025	0.9161	0.7633	0.3493	0.4549	0.6517	0.8362
W/3 Clusters	0.7731	0.2961	0.9091	0.7578	0.3423	0.4467	0.6429	0.8300
W/4 Clusters	0.7829	0.3112	**0.9510**	0.7641	**0.3596**	**0.4690**	**0.6739**	**0.8525**
W/5 Clusters	0.7731	0.2971	0.9161	0.7571	0.3435	0.4486	0.6466	0.8328
W/6 Clusters	0.7822	0.3079	0.9301	0.7657	0.3554	0.4626	0.6624	0.8439
W/7 Clusters	0.7759	0.3051	**<u>0.9580</u>**	0.7555	0.3533	0.4628	**0.6709**	**0.8508**
W/8 Clusters	0.7717	0.2957	0.9161	0.7555	0.3420	0.4471	0.6453	0.8319
W/9 Clusters	0.7822	0.3079	0.9301	0.7657	0.3554	0.4626	0.6624	0.8439
W/10 Clusters	0.7738	0.2995	0.9301	0.7563	0.3465	0.4532	0.6545	0.8387
W/11 Clusters	**0.7865**	**0.3113**	0.9231	**0.7712**	0.3589	0.4656	0.6627	0.8437
W/12 Clusters	0.7801	0.3093	**<u>0.9580</u>**	0.7602	0.3577	**0.4676**	0.6749	**<u>0.8534</u>**
W/13 Clusters	0.7801	0.3030	0.9091	0.7657	0.3497	0.4545	0.6494	0.8343
W/14 Clusters	**0.7900**	**0.3168**	**0.9371**	**0.7735**	**0.3651**	**0.4735**	0.6734	**0.8514**
W/15 Clusters	0.7674	0.2945	**0.9371**	0.7484	0.3413	0.4482	0.6524	0.8375

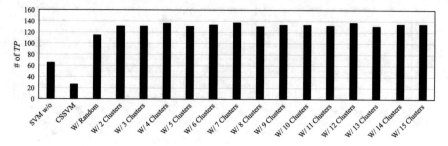

Fig. 4 Number of *TP* for methods

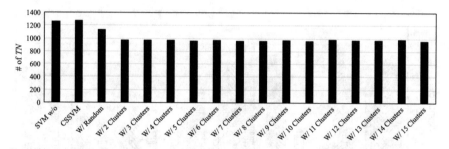

Fig. 5 Number of *TN* for methods

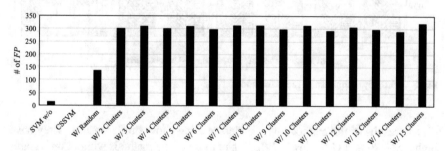

Fig. 6 Number of *FP* for methods

The SVM w/o method shows the best *ACC*; however it is second worst with respect to *SN*. This is because most outliers are classified into a class of normal values (the majority class) specified in Table 3, which is a common problem with imbalanced datasets.

The CSSVM method obtains a perfect result for *PREC*; however, it obtains the worst result for *SN*. This is because the method imposes heavier costs for misclassifying outliers and can reduce *FP*. At the same time, it increases *FN*. The results indicate that CSSVM can accurately detect some outliers; however it overlooks most outliers. In other words, the outliers predicted by CSSVM must be actual outliers; however, only a few outliers are detected by it.

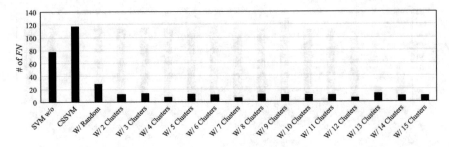

Fig. 7 Number of *FN* for methods

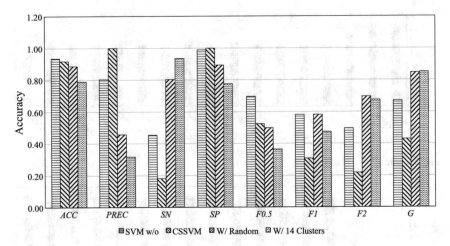

Fig. 8 Accuracy comparison of four methods

The results of the methods with undersampling show similar results. They have higher *SN*, F_2 and *G* but lower *PREC* than those of the methods without undersampling. The W/Randoms method obtains higher *PREC* than those of the methods with k-means cluster-based undersampling. In contrast, the W/*n* Clusters methods have better results in terms of *SN*. In addition, the classification criteria accuracy depends on the number of clusters, and all results of the W/14 Clusters method are within the top five. These results show that the proposed methods can improve the accuracy of detecting outliers; however they tend to classify too many samples as outliers.

6 Conclusion

This research examined the ability of undersampling to detect outliers in terms of the number of errors in embedded software development projects. The undersampling method was based on the k-means clustering algorithm. The method aimed at improving the proportion of outliers that were correctly identified while maintaining the other classification performance metrics high.

Evaluation experiments were conducted to compare the prediction accuracy of the methods with k-means undersampling, random undersampling and without undersampling using ten-fold cross-validation.

The results indicated that the methods with undersampling have higher sensitivity and lower precision than those of the methods without undersampling. The results further indicated that the proposed methods improved the accuracy of detecting outliers but classified too many samples as outliers.

In future research, we plan to investigate the following:

1. We plan to apply the oversampling method to improve the accuracy of precision, while keeping sensitivity high.
2. We intend to consider other methods to detect outliers.
3. More data are needed to further support our research. In particular, data for projects that include outliers are essential for improving the models.

Acknowledgements This work was supported by JSPS KAKENHI Grant Number JP16K00310 and JP17K00317.

References

1. Barandela, R., Sánchez, J.S., Garca, V., Rangel, E.: Strategies for learning in class imbalance problems. Pattern Recognit. **36**(3), 849–851 (2003). http://dblp.uni-trier.de/db/journals/pr/pr36.html
2. Boehm, B.: Software engineering. IEEE Trans. Softw. Eng. **C-25**(12), 1226–1241 (1976)
3. Fumera, G., Roli, F.: Cost-sensitive learning in support vector machines. In: VIII Convegno Associazione Italiana per L'Intelligenza Artificiale (2002)
4. Gordon, M., Kochen, M.: Recall-precision trade-off: a derivation. J. Am. Soc. Inf. Sci. **40**(3), 145–151 (1989)
5. Hartigan, J.A., Wong, M.A.: Algorithm as 136: A k-means clustering algorithm. J. R. Stat. Soc. Series C (Appl. Stat.) **28**(1), 100–108 (1979). http://www.jstor.org/stable/2346830
6. Hirayama, M.: Current state of embedded software (in japanese). J. Inf. Process. Soc. Jpn (IPSJ) **45**(7), 677–681 (2004)
7. Iwata, K., Liebman, E., Stone, P., Nakashima, T., Anan, Y., Ishii, N.: Bin-Based Estimation of the Amount of Effort for Embedded Software Development Projects with Support Vector Machines, pp. 157–169. Springer International Publishing (2016)
8. Iwata, K., Nakashima, T., Anan, Y., Ishii, N.: Error estimation models integrating previous models and using artificial neural networks for embedded software development projects. In: Proceedings of 20th IEEE International Conference on Tools with Artificial Intelligence, pp. 371–378 (2008)
9. Iwata, K., Nakashima, T., Anan, Y., Ishii, N.: Improving accuracy of an artificial neural network model to predict effort and errors in embedded software development projects. In: Lee, R., Ma, J., Bacon, L., Du, W., Petridis, M. (eds.) Software Engineering, Artificial Intelligence, Networking and Parallel/Distributed Computing 2010, *Studies in Computational Intelligence*, vol. 295, pp. 11–21. Springer Berlin Heidelberg (2010). doi:10.1007/978-3-642-13265-0_2
10. Iwata, K., Nakashima, T., Anan, Y., Ishii, N.: Estimating interval of the number of errors for embedded software development projects. Int. J. Softw. Innov. (IJSI) **2**(3), 40–50 (2014). doi:10.4018/ijsi.2014070104

11. Iwata, K., Nakashima, T., Anan, Y., Ishii, N.: Effort estimation for embedded software development projects by combining machine learning with classification. In: Proceedings of 3rd ACIS International Conference on Computational Science/Intelligence and Applied Information, pp. 265–270 (2016)
12. Komiyama, T.: Development of foundation for effective and efficient software process improvement. J. Inf. Process. Soc. Jpn (IPSJ) **44**(4), 341–347 (2003) (in japanese)
13. Masnadi-Shirazi, H., Vasconcelos, N.: Risk minimization, probability elicitation, and cost-sensitive svms. In: J. Fürnkranz, T. Joachims (eds.) ICML, pp. 759–766. Omnipress (2010). http://dblp.uni-trier.de/db/conf/icml/icml2010.html
14. Nakamoto, Y., Takada, H., Tamaru, K.: Current state and trend in embedded systems. J. Inf. Process. Soc. Jpn (IPSJ) **38**(10), 871–878 (1997) (in japanese)
15. Nakashima, S.: Introduction to model-checking of embedded software. J. Inf. Process. Soc. Jpn (IPSJ) **45**(7), 690–693 (2004) (in japanese)
16. Ogasawara, H., Kojima, S.: Process improvement activities that put importance on stay power. J. Inf. Process. Soc. Jpn (IPSJ) **44**(4), 334–340 (2003) (in japanese)
17. Sammut, C., Webb, G.I. (eds.): Encyclopedia of Machine Learning and Data Mining. 2. Springer US (2017)
18. Takagi, Y.: A case study of the success factor in large-scale software system development project. J. Inf. Process. Soc. Jpn (IPSJ) **44**(4), 348–356 (2003) (in japanese)
19. Tamaru, K.: Trends in software development platform for embedded systems. J. Inf. Process. Soc. Jpn (IPSJ) **45**(7), 699–703 (2004) (in japanese)
20. Ubayashi, N.: Modeling techniques for designing embedded software. J. Inf. Process. Soc. Jpn (IPSJ) **45**(7), 682–692 (2004) (in japanese)
21. Watanabe, H.: Product line technology for software development. J. Inf. Process. Soc. Jpn (IPSJ) **45**(7), 694–698 (2004) (in japanese)
22. Weiss, G.M.: Mining with rarity: a unifying framework. SIGKDD Explor. Newsl. **6**(1), 7–19 (2004). doi:10.1145/1007730.1007734

Development of Congestion State Guiding System for University Cafeteria

Takafumi Doi, Hirotaka Ito and Kenji Funahashi

Abstract We develop a congestion state guide system for a university cafeteria. This system confirms the congestion state of a cafeteria using iBeacon, and displays it on a mobile device in real time. The purpose of this is to stagger cafeteria use and avoid rush hour. We had a field experiment of this system in a cafeteria, and tested the operation of this system. Using this system, students can know how congested a cafeteria is at any place outside the cafeteria, and have a lunch comfortably when it was not crowded.

Keywords Bluetooth low energy · iBeacon · Congestion state guide · University cafeteria · Staggered lunch time

1 Introduction

In recent years, smartphones have become popular, and services for a smartphone and systems using its location information are widely used. Especially outdoor and wide area location data are utilized that uses GPS, the Internet access point information etc., and many applications are developed [1, 2]. However the number of applications using indoor and local area location data are not so many, because it is difficult to receive GPS signal, and a device usually moves inside one access point area. One of the indoor positioning method estimates position information using beacons that transmit radio waves. WSN (Wireless Sensor Networks) is also similar technology [3, 4]. There is a indoor positioning system study using BLE beacon [5]. In order to improve the accuracy in this study, authors combined a plurality of methods. For example, there is also indoor localization method and open field trial [6].

T. Doi · H. Ito · K. Funahashi (✉)
Nagoya Institute of Technology, Gokiso-cho, Showa-ku, Nagoya 466-8555, Japan
e-mail: kenji@nitech.ac.jp

T. Doi
e-mail: taka@center.nitech.ac.jp

H. Ito
e-mail: ht-itoh@nitech.ac.jp

© Springer International Publishing AG 2018   81
R. Lee (ed.), *Computational Science/Intelligence and Applied Informatics*,
Studies in Computational Intelligence 726, DOI 10.1007/978-3-319-63618-4_7

While indoor positioning systems are gaining attention, iBeacon [7] was proposed by Apple in September 2013. It uses Bluetooth Low Energy (BLE) [8] technology, and smartphone and tablet receive BLE signals. These devices can detect entering into and leaving from beacon area.

Nagoya Institute of Technology has introduced a system to manage attendance of students using ID card and a smart card reader to register their arrival at class. An attendance register subsystem using BLE beacon and their own smartphone has also introduced recently. It is expected that university life become more convenient.

By the way, many university cafeterias have a same problem that most students want to have a lunch at a similar time, and a cafeterias are crowded very much. But it is difficult to enlarge cafeteria space because the number of students is limited and just most students concentrate there between the classes. Although simple solution is to avoid the peak time to use and some students have a lunch before or after designated lunch time, it is difficult to predict the peak because university class schedule with canceled and supplemented classes is unstable. Therefore we expect that they will stagger cafeteria use and avoid rush hour by themselves if they can know the congestion state of a cafeteria before coming there from a class room and laboratory. Introducing internet web cameras there, they probably can know the state if it is crowded. But it needs much cost, and the cameras are only for this purpose. We consider to develop a congestion state guide system for a university cafeteria using our university BLE beacon system also attached at a cafeteria. Aside to that this beacon system is not only for attendance management, but also a guid system for visually impaired person, disaster evacuation guidance, etc. This system informs the congestion situation to students outside a cafeteria in real time. We had a field experiment of this system in a cafeteria, and tested the operation of this system. Using this system, students can know how congested a cafeteria is at any place outside the cafeteria, and have a lunch comfortably when it was not crowded.

In Sect. 2, we explain BLE and iBeacon used in our system. In Sect. 3, we describe the configuration of our system proposed in this paper. In Sect. 4, we describe an experiment and a result. In Sect. 5, we describe conclusion and future works.

2   Bluetooth Technology

2.1   Bluetooth Low Energy

In this section, we explain Bluetooth Low Energy (BLE) technology to help to understand our guiding system. BLE is one of Bluetooth standards. BLE is formulated by the Bluetooth special interest group (SIG). BLE is installed after the Bluetooth 4.0 standard, and it is possible to operate with low power. BLE is a specification derived from Wibree which is a proximity communication technology developed by Nokia as its own wireless standard, so it is not compatible with Bluetooth 3.0 or earlier.

Therefore, terminals equipped with Bluetooth 4.0 or later are often implemented so that conventional Bluetooth 3.0 can also be used.

Frequency band is used by dividing into 40 channels of 2 MHz width at the 2.4 GHz band. Three channels are allocated to an advertisement channel used for discovery and connection of BLE devices, and the rest channels are allocated to data channels used for data communication. Frequency hopping is used as a method to avoid interference. As a result, even if a specific channel can not communicate, data communication can be continued if the channel is switched after a certain time.

Broadcast and connection are defined in BLE as a method between devices for communication. Broadcast is a communication method for unilaterally transmitting data from one BLE device to other devices. In this communication method, a device that transmits data is called a broadcaster, and a device that receives data is called an observer. It is a characteristic that a broadcaster transmits the same data to unspecified observers at the same time. Connection is a communication method for mutually transmitting and receiving data between a BLE device and another BLE device. Transmission and reception of data is private only between the connected devices.

2.2 iBeacon

We explain also iBeacon in this section. iBeacon is a technology for position detection and proximity detection using BLE proposed by Apple. This technology uses broadcast communication of BLE. The sending terminal is called the peripheral, the receiving terminal is called the central, and advertising that the peripheral sends out information is called the advertisement. Central receives radio waves advertised from peripherals and central processes according to the radio content.

iBeacon has two major functions. These are to detect entering and leaving from the area and to detect the extent of proximity between the peripheral terminal and the central terminal. The entrance and exit of the area is made based on whether or not the radio wave emitted by the peripheral terminal has arrived. For the detection of the degree of proximity, we estimate the approximate distance using RSSI. iBeacon is compatible with iOS 7.0 or later, Android 4.3 or later.

3 System

3.1 Client Application

To guide the congestion state of a cafeteria, the server to summarize it is needed. The client application is also necessary for users to know the summarized data. Therefore, our proposed system consists of client application and server software. The client application should have two functions. One of them is to detect entering into and leaving from the cafeteria, and then send the information to the server. Another is to fetch the congestion state of the cafeteria, and then display it on the client mobile device monitor.

In order to detect entering into and leaving from the cafeteria, it is necessary to detect iBeacon. We use AltBeacon library published by Radius Networks [9]. When a client device detects a beacon attached in a cafeteria, the detection process for entering into beacon area is performed, and then the proximity detection of a beacon is performed. Because the distance to the detected beacon can not be known when it detects entering into the beacon area. A client device should judge whether it is inside or outside a cafeteria by using the proximity detection of beacons. Figure 1 shows the situation when the client device enters into a cafeteria. When it detects some beacons, and the nearest beacon is at a cafeteria, it judges that it is inside a cafeteria. Figures 2 and 3 show the situation when the client device leaves from the cafeteria. When it detects some beacons too, and the nearest beacon is not at a cafeteria, it judges that it is outside a cafeteria. And when no beacons are detected, it judges that it is also outside a cafeteria.

The distance to a beacon is used as a criterion of this proximity detection. The distance is obtained from Transmission Power (TxPower) and Received Signal Strength Indication (RSSI) as following equations.

$$A = \frac{TxPower - RSSI}{20} \tag{1}$$

$$distance = 10^A \tag{2}$$

The client application judges whether it is inside or outside a cafeteria at 20 s intervals. When the client application confirms entering into or leaving from the cafeteria, it is necessary to send the information to the server. In addition, it also notifies the action to a user for confirmation.

Figures 4 and 5 show an example of the application screen. The client displays five items on the screen, and has one button for updating information. The five items are:

- Information on whether the user is inside or outside a cafeteria
- The number of system users in a cafeteria
- The number of people who use cafeteria estimated from the number of system users in cafeteria
- Three levels evaluation of congestion state
- The transition of the estimated number of people in a cafeteria in the past

Figure 6 shows the process of this system when the button is pressed. The client application sends a request to the server, receives a response from the server, and displays each data.

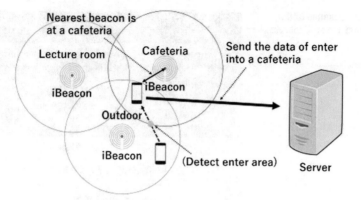

Fig. 1 Situation that client device enters into a cafeteria

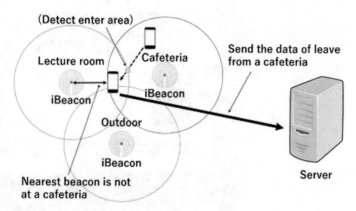

Fig. 2 Situation that client device leaves from the cafeteria

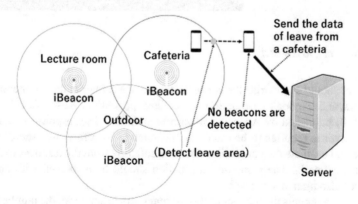

Fig. 3 Situation that client device leaves from the cafeteria and any

Fig. 4 An example of the
application screen (Japanese
version)

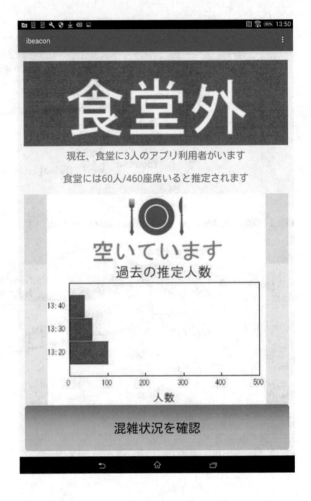

3.2　Server Software

The server should summarize the congestion state of a cafeteria; the number of cafeteria users, and respond for a request from a client application. We use Apache HTTP Server [10] published by Apache Software Foundation. The response information from a server is requires to be easier to understand. Therefore, this server should estimate and evaluate the congestion state of a cafeteria from the turnover of system users at a cafeteria. The response information should be displayed with not only letters but also figures and graphs.

When a server gets the number of this system users in cafeteria, the number of the actual students having lunch there should be considered. So it estimates the number of student who are in a cafeteria first, and then judges the congestion state in three levels. It also serves the transition of the estimated number of people in the last 30 min

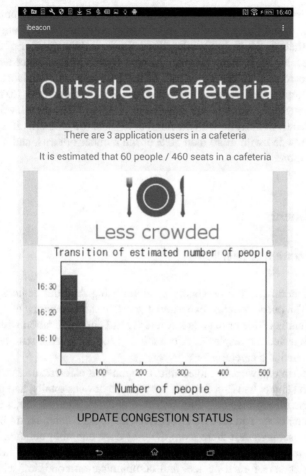

Fig. 5 An example of the application screen (English version)

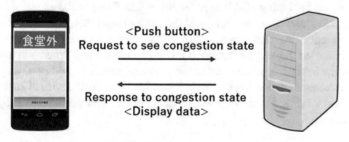

Fig. 6 Process of this system when the button is pressed

at 10 min intervals. The actual number of people there is estimated from both of the number of the system students using the attendance register subsystem using BLE beacon with their own smartphone described above, and the number of the students having classes, it means, using the management system of attendance using ID card and a smart card reader. The usage rate of the BLE beacon attendance subsystem is 5% for example, the estimated number of people in a cafeteria is 20 times of the number of our system users who are detected in a cafeteria. The three level congestion state is decided from the ratio with the number of seats in a cafeteria. It is judged that it is a little crowded with more than 50% of the number of seats, and it is crowded with 80% or more.

4 Experiment

4.1 Experimental Method

We had an experiment at a university cafeteria using Android devices with client application. The purpose of this experiment was to test the accuracy of the judgment whether a client is inside or outside a cafeteria, and communication with the server. In addition, it tested the displayed information of the congestion state of a cafeteria at any place outside a cafeteria.

In this experiment, we asked five subjects to move as cafeteria users with Android devices. Each subject had 2–5 devices, we used 18 devices totally, and they entered into a cafeteria through the east door one after another. Figure 7 shows the iBeacon positions, an entrance and an exit. No devices left from a cafeteria until all devices entered. Then subjects left there from the east door one after another. No devices re-entered. We recorded the actual number of the devices in a cafeteria, beacon confirmation status of each device, and communication condition with the serve at every time of device entering and leaving. Each recording time is numbered in order from the beginning of this experiment. Figure 8 shows the appearance of this experiment at a cafeteria. Subjects held the devices with their hand or put them on a table in a cafeteria. After entering into a cafeteria, subjects sit apart there. At that time we recorded the beacon detection result of each device. When leaving from a cafeteria, the beacon detection results were also recorded. In addition, we checked the server communication log of sent information from each device at each time of entering and leaving. Figure 9 shows the client device that displayed a subject was outside cafeteria.

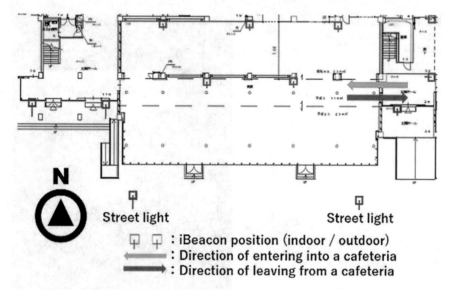

Street light　　　　　　　　　　　　　**Street light**

⊓ ⊓ : iBeacon position (indoor / outdoor)
⟵ : Direction of entering into a cafeteria
➡ : Direction of leaving from a cafeteria

Fig. 7 Entrance, exit and iBeacons

Fig. 8 Appearance of
experiment at a cafeteria

4.2 Result and Consideration

First we confirmed that the congestion state could be shown on the client screen
outside a cafeteria from Fig. 9. Experimental results of entering into and leaving
from a cafeteria are shown in Tables 1 and 2. We used all 18 device data at the
entering, and for leaving process we used 15 data of devices that were detected
correctly when entering.

Table 1 shows experimental results of entering into a cafeteria. The 16 client
devices judged their entering correctly. This means 88% accuracy. We guess that the

Fig. 9 Appearance of
experiment outside a
cafeteria

miss-judging devices were alongside the wall, then the RSSI of the beacon outside a
cafeteria was strong. The 10 client devices sent the entering information to the server
correctly. This means 55% of all 18 devices, and 62% of the 16 devices that judged
their entering correctly. This communication result was far from satisfactory. We
consider the reason is that the client application tried to communicate with the server
to send the information, before the client woke up from the Wi-Fi sleep completely.

Table 2 shows experimental results of leaving from a cafeteria. The devices that
judged leaving from a cafeteria correctly were all devices that judged entering there.
This result indicates that entering judgement is more difficult than leaving judgement,
especially when a client is alongside the wall, and close to other beacon outside a
cafeteria. The 12 client devices sent the leaving information to the server correctly.
This means 80% of the target 15 devices, and it was higher than the communication
result of the entering. We guess that most subjects left from there before the devices
slept down with Wi-Fi.

The client application almost could detect an entering into and leaving from a
cafeteria at this field experiment. The comparison of the distance between a device
and a detected beacon also worked well to judge whether the device was inside or
outside a cafeteria. However, some clients could not send an entering information to
a server. It is necessary to solve this problem because the reliability of a congestion
state would become wrong. A client should send the data after the confirmation of
Wi-Fi communication, and should confirm the acknowledgement from a server, then
retry to send it again if necessary. An exact judgement is important of course, and an
exact data transfer is also important to make a guidance of congestion state better.

Table 1 Results of experiment when entering into a cafeteria

Recording time number	1	2	3	4	5	6
Total number of devices entered cafeteria	2	7	10	12	15	18
Number of devices judged an entering correctly	2	6	9	11	13	16
Number of devices sent an entering data	2	4	5	8	9	10

Table 2 Results of experiment when leaving from a cafeteria

Recording time number	7	8	9	10	11
Total number of devices left cafeteria	3	5	10	12	15
Number of devices judged a leaving correctly	3	5	10	12	15
Number of devices sent an leaving data	3	5	8	9	12

5 Conclusion

Our final goal is to make a cafeteria more comfortable without waiting time for having lunch. In this paper, we proposed a new guiding system of a congestion state for a university cafeteria using our university BLE beacon system. We built the system that consists of client application and server software. The number of people having a lunch in a cafeteria can be counted, then the congestion state of a cafeteria can be provided onto a mobile device.

We also had an experiment for our proposed system, then the proximity detection of iBeacon worked well to accurately judge whether a client is inside or outside a cafeteria. And the system user could know the congestion state of a cafeteria also outside a cafeteria. Although clients could not communicate with a server to send entering and leaving information of client because of Wi-Fi sleep probably, the experiment result was, on the whole, satisfactory.

For future work, we should consider that a client tries to send entering and leaving information to a server after waking up from Wi-Fi sleep, and a server accurately count the number of people in a cafeteria. After improvement this, we should have an experiment again. We also would like to rebuild an application to save power consumption. The client application described in this paper is only for Android OS. Therefor it is necessary to build an iOS application. Finally we will have a field demonstration, we expect that students will stagger cafeteria use and avoid rush hour by themselves after getting the congestion state of a cafeteria before coming there from a class room and laboratory.

Acknowledgements The authors would like to thank our colleagues in our laboratory for useful discussions. The experiment of this work was supported in part by NTT DOCOMO, INC.

References

1. Al-Naima, F.M., Ali, R.S., Abid, A.J.: Design of an embedded solar tracking system based on GPS and astronomical equations. Int. J. Inf. Technol. Web Eng. **9**(1), 12–30 (2014)
2. Zhou, B., Hsu, J., Wang, Y.: GIS and GPS applications in emerging economies: observation and analysis of a chinese logistics firm. Int. J. Inf. Syst. Soc. Change **1**(3), 45–61 (2010)
3. Gao, R., Couch, A., Chang, C.H.: An innovative positioning system with unsynchronized interferometric modulated signals in wireless sensor networks. Int. J. Networked Distrib. Comput. **5**(2), 101–112 (2017)
4. Zhang, Z.: Modeling performance of CSMA/CA with retransmissions in wireless personal area networks. Int. J. Networked Distrib. Comput. **1**(2), 97–107 (2013)
5. Kudou, D., Horikawa, M., Furudate, T., Okamoto, A.: The proposal of indoor positioning system by area estimation using BLE beacon. In: The 78th National Convention of Information Processing Society of Japan (IPSJ), vol. 3, pp. 245–426 (2016) (in Japanese)
6. Ishizuka, H., Kamisaka, D., Kurokawa, M., Watanabe, T., Muramatsu, S., Ono, C.: A fundamental study on a indoor localization method using BLE signals and PDR for a smart phone: Sharing results of exmeriments in open beacon field trial, IEICE (The Institute of Electronics, Information and Communication Engineers). Technical Report MoNA (Mobile network and applications), vol. 114, no. 31, pp. 133–138 (2014) (in Japanese)
7. Apple Inc.: iBeacon for Developers. https://developer.apple.com/ibeacon/
8. Bluetooth SIG: Bluetooth Low Energy. https://www.bluetooth.com/what-is-bluetooth-technology/how-it-works/low-energy
9. Radius Networks, Inc.: AltBeacon. http://altbeacon.org/
10. Apache Software Foundation: Apache HTTP Sever. https://httpd.apache.org/

Analog Learning Neural Circuit with Switched Capacitor and the Design of Deep Learning Model

Masashi Kawaguchi, Naohiro Ishii and Masayoshi Umeno

Abstract In the neural network field, many application models have been proposed. Previous analog neural network models were composed of the operational amplifier and fixed resistance. It is difficult to change the connecting weight of network. In this study, we used analog electronic multiple and switched capacitor circuits. The connecting weights describe the input voltage. It is easy to change the connection coefficient. This model works only on analog electronic circuits. It can finish the learning process in a very short time and this model will enable more flexible learning. However, the structure of this model is only one input and one output network. We improved the number of unit and network layers. Moreover, we suggest the possibility of the realization of the hardware implementation of the deep learning model.

Keywords Electronic circuit · Neural network · Multiple circuit · Deep learning

1 Introduction

We propose the dynamic learning of the neural network by analog electronic circuits. This model will develop a new signal device with the analog neural electronic circuit. One of the targets of this research is the modelling of biomedical neural function.

M. Kawaguchi (✉)
Department of Electrical & Electronic Engineering, Suzuka National College of Technology, Shiroko, Suzuka Mie, Japan
e-mail: masashi@elec.suzuka-ct.ac.jp

N. Ishii
Department of Information Science, Aichi Institute of Technology, Yachigusa, Yagusa-cho, Toyota, Japan
e-mail: ishii@aitech.ac.jp

M. Umeno
Department of Electronic Engineering, Chubu University, 1200 Matsumoto-cho, Kasugai, Aichi 487-8501, Japan
e-mail: umeno@isc.chubu.ac.jp

© Springer International Publishing AG 2018   93
R. Lee (ed.), *Computational Science/Intelligence and Applied Informatics*,
Studies in Computational Intelligence 726, DOI 10.1007/978-3-319-63618-4_8

In the field of neural network, many application models have been proposed. And there are many hardware models that have been realized. These analog neural network models were composed of the operational amplifier and fixed resistance. It is difficult to change the connection coefficient.

1.1 Analog Neural Network

The analog neural network expresses the voltage, current or charge by a continuous quantity. The main merit is it can construct a continuous time system as well as a discrete time system by the clock operation. Obviously, the operation of the actual neuron cell utilizes analog. It is suitable to use an analog method for imitating the operation of an actual neuron cell. Many artificial neural networks LSI were designed by the analog method. Many processing units can be installed on a single-chip, because each unit can be achieved with a small number of elements, addition, multiplication, and the nonlinear transformation. And it is possible to operate using the super parallel calculation. As a result, the high-speed offers an advantage compared to the digital neural network method [1, 2]. In the pure analog circuit, the main problem is the achievement of an analog memory, how to memorize analog quantity [3]. This problem has not been solved yet. The DRAM method memorizes in the capacitor as temporary memory, because it can be achieved in the general-purpose CMOS process [4]. However, when the data value keeps for a long term, digital memory will also be needed. In this case, D/A and A/D conversion causes an overhead problem. Other memorizing methods are the floatage gate type device, ferroelectric memory (FeRAM) and magnetic substance memories (MRAM) [5, 6].

1.2 Pulsed Neural Network

Another hardware neural network model has been proposed. It uses a pulsed neural network. Especially, when processing time series data, the pulsed neural network model has good advantages. In particular, this network can keep the connecting weights after the learning process [7]. Moreover, the reason the learning circuit used the capacitor is that it takes a long time to work the circuits. In general, the pulse interval of the pulsed neural network is about 10 μS. The pulsed neuron model represents the output value by the probability of neuron fires. For example, if the neuron is fired 50 times in a 100 pulse interval, the output value is 0.5 at this time. To represent the analog quantity using the Pulsed Neuron Model, there needs to be about 100 pulses. Thus, about 1mS is needed to represent the output analog signal on a pulsed neuron model.

In this study, we used multiple circuits. The connecting weights describe the input voltage. It is easy to change the connection coefficient. This model works only on analog electronic circuits. It can finish the learning process in a very short time and this model will allow for more flexible learning. Recently, many researchers have focused on the semiconductor integration industry. Especially, low electrical power, low price, and large scale models are important. The neural network model explains the biomedical neural system. Neural network has flexible learning ability. Many researchers simulated the structure of the biomedical brain neuron using an electronic circuit and software.

1.3 Overview

The results of the neural network research provide feedback to the neuro science fields. These research fields were have been widely developed. The learning ability of a neural network is similar to the human mechanism. As a result, it is possible to make a better information processing system, matching both advantages of the computer model and biomedical brain model. The structure of the neural network usually consists of three layers, the input layer, intermediate layer and output layer. Each layer is composed of the connecting weight and unit. A neural network is composed of those three layers by combining the neuron structures [8, 9].

In the field of neural network, many application methods and hardware models have been proposed. A neuro chip and an artificial retina chip are developed to comprise the neural network model and simulate the biomedical vision system. In this research, we are adding the circuit of the operational amplifier. The connecting weight shows the input voltage of adding circuits. In the previous hardware models of neural net-work, changing connected weights was difficult, because these models used the resistance elements as the connecting weights.

Moreover, the model which used the capacitor as the connecting weights was proposed. However, it is difficult to adjust the connecting weights. In the present study, we proposed a neural network using analog multiple circuits. The connecting weights are shown as a voltage of multiple circuits. The connecting weights can be changed easily. The learning process will be quicker. At first we made a neural network by computer program and neural circuit by SPICE simulation. SPICE means the Electric circuit simulator as shown in the next chapter. Next we measured the behavior confirmation of the computer calculation and SPICE simulation. We compared both output results and confirmed some extent of EX-OR behavior [10, 11].

2 SPICE

In this research, we used the electric circuit simulator SPICE. Electric circuit simulator (SPICE) is the abbreviation of Simulation Program with Integrated Circle Emphasis. It can reproduce the analog operation of an electrical circuit and the elec-

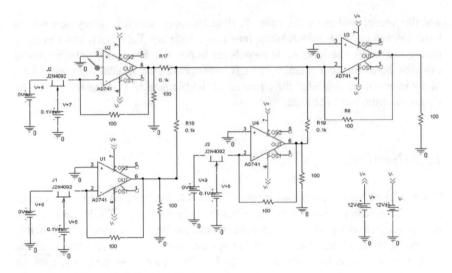

Fig. 1 Neural circuit (Two-input and One-output)

tric circuit. After this, the circuit drawn by CAD, set the input voltage. SPICE has the function of AC, DC and transient analysis. At first, we made the differential ampli-fier circuits and Gilbert multipliers circuits. And we confirmed the range of voltage operated excellently. The neuron structure was composed of multiple circuits by an operational amplifier for multiplication function achievement, current mirror circuits to achieve nonlinear function and differential amplifier circuits.

In the previous hardware model of neural network, we used the resistance element as a connecting weight. However, it is difficult to change the resistance value. In the neural connection, it calculates the product the input value and connecting weight. We used the multiple circuit as the connecting weight. Each two inputs of multiple circuits means an input value and connecting weight. The connecting weight shows the voltage value. It is easy to change the value in the learning stage of neural network. Figure 1 is the neural circuit of two inputs and one output which reproduces the characteristic of one neuron, using current addition by current mirror circuits, the product of the input signal and connecting weights. In the neural circuit, we have to use adder and subtraction circuit as input part of each synapse. In our previous research, we used adder and subtraction circuit by operational amplifier.

However in this research, we used Switched Capacitor Adder/subtracter circuit. This circuit can hold the output signal in a brief time. It is useful when neural network is on working stage after finished learning.

Fig. 2 Switched Capacitor
circuits

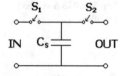

Fig. 3 MOS Switched
circuits

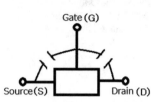

Fig. 4 MOS Switched
circuits

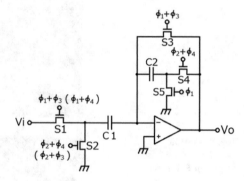

3   Switched Capacitor Circuit

Switched Capacitor circuits are composed of switches and capacitors. The charge
is moved by switching. Its circuits have the function of electrical resistance by fast
switching as Fig. 2. We can make these Switched Capacitor circuits using MOS
switch as Fig. 3. So it is easy to make the integrated circuits.

3.1   Switched Capacitor Adder/Subtracter

The circuit shown in Fig. 4 is the configuration of the switched capacitor adder/
subtracter circuits. Figure 5 is the output waveform of the switched capacitor adder-
subtracter circuits. The previous model uses the resistance for adder/subtracter circuits. It is possibility to generate favor and spend more energy. On the other hand,
the capacitor does not generate heat during operation.

This circuit operates with 4 phase clocks, Φ_1, Φ_2, Φ_3, Φ_4, without overlapping.
When the two clocks, Φ_1 and Φ_2 are generated, these circuits operate as a non-
inverting amplifier. When the clock Φ_1 is generated, V_1 is accumulated in the capacitor. On the other hand, when the two clocks, Φ_3 and Φ_4 are generated, these circuits

Fig. 5 MOS Switched circuits

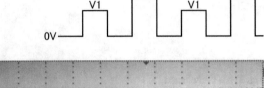

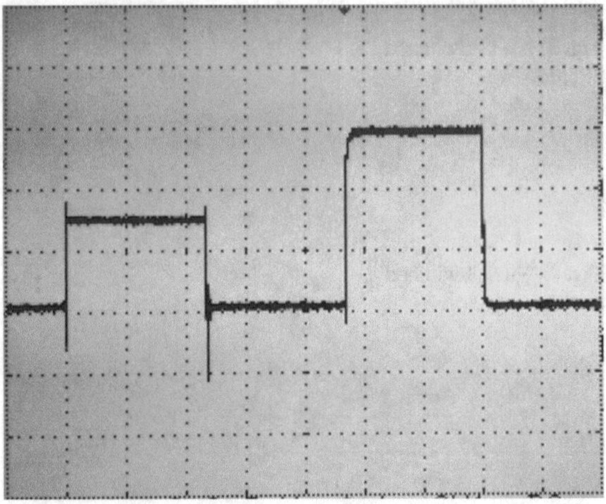

Fig. 6 The Output waveform of Switched Capacitor Adder

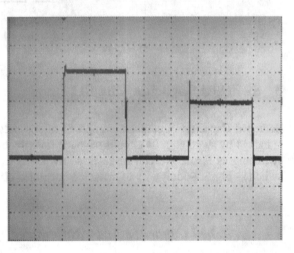

Fig. 7 The Output waveform of Switched Capacitor Subtracter

operate as an inverting amplifier. When the clock Φ_3 is generated, $V_1 + V_2$ is accumulated in the capacitor.

Further, this circuit can operate as a subtracter circuit when the clock signals, Φ_1 and Φ_4, applied to the switch S_1 and Φ_2 and Φ_3, applied to the switch S_2 in Fig. 4 as shown in parentheses. When the clock Φ_1 is generated, V_1 is accumulated in the capacitor. On the other hand, when the clock Φ_3 is generated, V_2 is put out from the capacitor. This circuit operates as a subtracter, the output signal is $V_1 - V_2$.

3.2 The Experiment of Switched Capacitor Adder/Subtracter Circuit

Figure 6 shows the Output waveform of Switched Capacitor Adder. The input voltage V_1 is 3.0 [V] and V_2 is 3.0 [V]. The first clock means $V_1 = 3.0$ [V] and the second clock means $V_1 + V_2 = 6.0$ [V]. This circuit operates as an Adder. Figure 7 shows the Output waveform of Switched Capacitor Subtracter. The input voltage V_1 is 6.0 [V] and V_2 is 3.0 [V]. The first clock means $V_1 = 6.0$ [V] and the second clock means $V_1 + V_2 = 3.0$ [V]. This circuit operates as a Subtracter.

3.3 A/D Converter Circuits

These proposed circuits have pure analog elements. However, in the practical use, the digital signal will be required. It needs the A/D converter circuits. It converts analog signal to digital signal. We show the A/D converter circuits using switched capacitor by double integrating type in Fig. 8. We used this switched capacitor adder-subtracter circuits in this learning Neural network. Figure 9 shows the output waveform of Switched Capacitor A/D converter and Fig. 10 shows the output waveform of NN circuit.

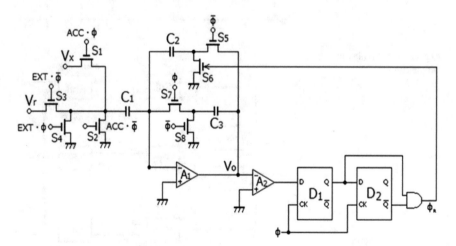

Fig. 8 The circuit of Switched Capacitor A/D converter

Fig. 9 The output waveform of Switched Capacitor A/D converter

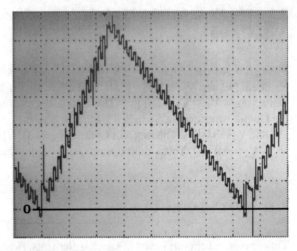

Fig. 10 The Output waveform of NN circuit

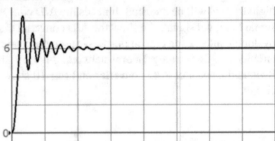

Fig. 11 The Architecture of Perceptron

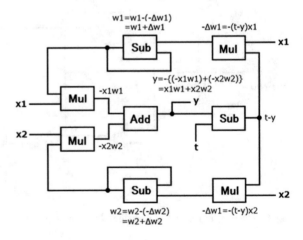

4 Perceptron Network by Analog Circuits

At first, we constructed a two-input and one-output perceptron neural network. In Fig. 11, we show the block diagram of perceptron neural network. "Mul" means multiple circuits, "Add" means addition of circuits and "Sub" means Subtraction circuits in Fig. 11. Figure 12 shows the perceptron circuits, two-input and one-output. The learning time is about 900 μS [12] Fig. 13 shows the architecture of three-layers neural circuits and Fig. 14 shows the learning neural circuit on capture CAD by SPICE.

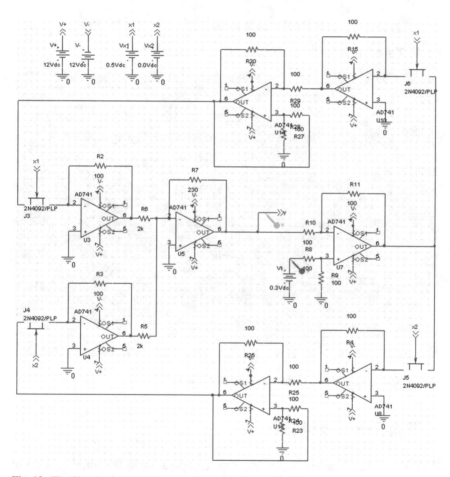

Fig. 12 The Circuit of perceptron

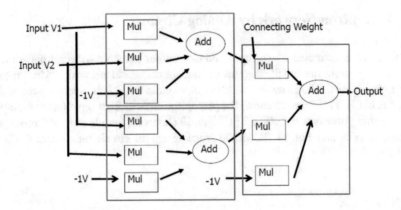

Fig. 13 The architecture of three-layers neural circuits

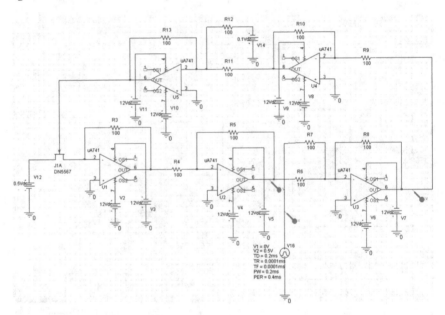

Fig. 14 The Learning Neural Circuit on Capture CAD by SPICE

5   Deep Learning Model

Recently, a deep learning model has been proposed. Deep learning is a kind of algorithms in learning model. It attempts the high-level categorizing of data using multiple non-linear transformations and one method of machine learning. In the field of image recognition and speech recognition, the deep learning method has attracted the attention.

5.1 The Stacked Auto Encoder

The stacked auto-encoder is one method of deep learning. This is the pre-learning method of large number layer network. How to construct the deep layer network is as follows.

After the learning process of stacked auto-encoder is completed, remove the decoding part (output layer) of stacked auto-encoder and keep the coded portion (from the input layer to the intermediate layer). Thus we obtain the network which converts from input signal to compressed information representation shown in Fig. 15.

Moreover, we obtain more compressed internal representation, as the compressed representation input signal to apply the auto-encoder learning shown in Fig. 16. Thus, we obtain a multi-layered hierarchical network, recursively repeated auto-encoder learning, and stacked the encoding part of the network. This constructed multilayer network is called stacked auto-encoder. In this way, after building a multi-layer network, to add the identified network using the output of the final layer, a new supervised learning method is proposed.

Stacked auto-encoder has been applied to the various subjects as well as the DNN which is stacked the RBM. Recently, the learning experiment of feature extractor from a large amount of image has become famous.

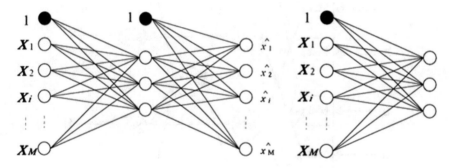

Fig. 15 Learning the Auto-encoder and removing the decoding part of stacked auto-encoder

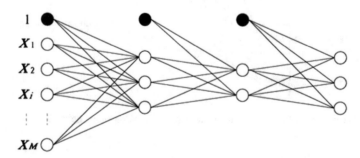

Fig. 16 The compressed internal representation

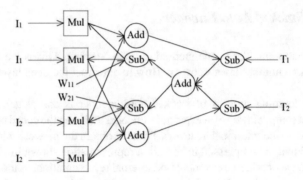

Fig. 17 The structure of 2-input, 1-output and 2-patterns analog neural network

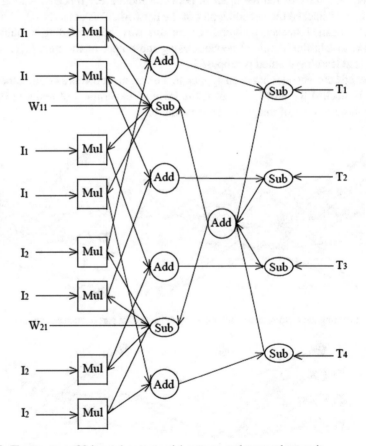

Fig. 18 The structure of 2-input, 1-output and 4-patterns analog neural network

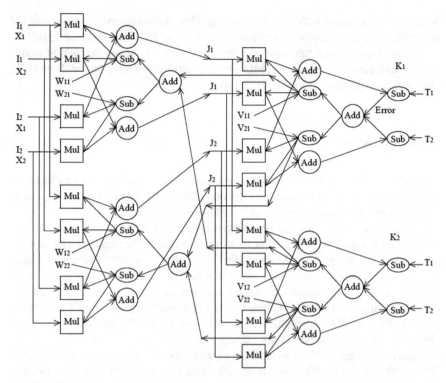

Fig. 19 The structure of 2-input, 1-output, 2-patterns and 3-layers analog neural network

This network structure is nine layers with three superimposed sub-network such as convolution network. It network learned 10 million images as an input signal which are cut out from 10 million pieces of YouTube video. Each sub-network is made by unsupervised learning using the stacked auto-encoder method. As a result, it has been reported that the neurons are formed which can respond specifically to various kinds of objects types such as a human face, cat face and drink bottles [13].

In the previous research, we described the dynamical neural network learning model. However, this model has only one input unit and one output unit. To realize the hardware deep learning model, we have to increase the number of units in each layer. Next, we constructed a 2 input, 1 output and 2 patterns neural model as in Fig. 17. I_1 and I_2 are input units. Two I_1 mean two inputs. T_1 and T_2 means two teaching signals. W_{11} and W_{12} are connecting weights. Figure 18 means the structure of 2-input, 1-output and 4-patterns analog neural network. It means 4 input and 4 teaching signals of each pattern. Figure 19 means the structure of 2-input, 2-output, 2-patterns and 3-layers analog neural network. Although this model needs many neural connections, the learning speed is very high. After learning, picking up each new connecting weights between input layer and middle layer. These connecting weights are used by the one layer of the deep learning model. We will pick up each output value of the middle layer. These output values of the middle layer uses as the input value of next repeat of auto-encoder learning. This model suggests the possibility of the realization of the hardware implementation of the deep learning model [14].

We used operational amplifier as a representation of connecting weight in this experiment. However when using MOSFET circuits instead of operational amplifier, the whole size of this model will be reduced.

6 Conclusion

We constructed a three layer neural network, with two-input layers, two-middle layers and one output layer. We confirmed the operation of the three layer analog neural network with the multiplying circuit by SPICE simulation.

The connection weight can be changed by controlling the input voltage. This model has extremely high flexibility characteristics. When the analog neural network is operated, the synapse weight is especially important. It is how to give the synapse weight to this neural network. To solve this problem, it is necessary to apply the method of the back propagation rule that is a general learning rule for the multiple electronic circuits. This neural circuit model is possible the learning. The learning speed will be rapid. And dynamic learning will be realized. The method is calculating the difference between the output voltage and the teaching signal of the different circuits and the feedback of the difference value for changing connecting weights. The learning cycle of this circuit is 25,000 times per second. The learning speed of this model is very high in spite of a very simple circuit using low cost elements.

The learning time of this model is very short and the working time of this model is almost real-time. Because this model can learn different patterns in the same time. The pulsed neuron model represents the output value by the probability of neuron fires. To represent the analog quantity using the Pulsed Neuron Model, enough time for at least a few dozen pulses is needed. The output value of this model is the output voltage of this circuit. We don't need to convert the data; we can use the raw data from this model. This model allows for switching the working mode and learning mode. It is always necessary to input the teaching signal. However, the connecting weight changes according to the changing of the teaching signal. This model can also easily accommodate changes in the environment. In each scene, optimal learning is possible. It will improve the artificial intelligence element with self dynamical learning. The realization of an integration device will enable the number of elements to be reduced. The proposed model is robust with respect to fault tolerance. Future tasks include system construction and mounting a large-scale integration.

Moreover, deep learning method is recently proposed. If this system improved toward the deep learning model, many applications will be realized. It is a kind of algorithms in the learning model. It attempts making high-level categorizing data using multiple non-linear transformations and one method of machine learning. In the field of image recognition and speech recognition, the deep learning method has attracted the attention. We suggested the possibility of the realization of the hardware implementation of the deep learning model. It will improve the artificial intelligence element with self-dynamical learning. The realization of an integration device will enable the learning time to be reduced. The proposed model is robust with respect to fault tolerance. Future tasks include system construction and mounting a large-scale integration.

References

1. Mead, C.: Analog VLSI and Neural Systems. Addison Wesley Publishing Company, Inc. (1989)
2. Chong, C.P., Salama, C.A.T., Smith, K.C.: Image-motion detection using analog VLSI. IEEE J. Solid-State Circuits $27(1)$, 93–96 (1992)
3. Lu, Z., Shi, B.E.: Subpixel resolution binocular visual tracking using analog VLSI vision sensors. IEEE Trans. Circuits Syst.-II: Analog Digit. Signal Process. $47(12)$, 1468–1475 (2000)
4. Saito, T., Inamura, H.: Analysis of a simple A/D converter with a trapping window. IEEE Int. Symp. Circuits Syst., 1293–1305 (2003)
5. Luthon, F., Dragomirescu, D.: A cellular analog network for MRF-based video motion detection. IEEE Trans. Circuits Syst.-I: Fund. Theory Appl. $46(2)$, 281–293 (1999)
6. Yamada, H., Miyashita, T., Ohtani, M., Yonezu, H.: An analog MOS circuit inspired by an inner retina for producing signals of moving edges. Technical Report of IEICE, NC99-112 , pp. 149-155 (2000)
7. Okuda, T., Doki, S., Ishida, M.: Realization of back propagation learning for pulsed neural networks based on delta-sigma modulation and its hardware implementation. ICICE Trans. J88-D-II-4, 778–788 (2005)
8. Kawaguchi, M., Jimbo, T., Umeno, M.: Motion detecting artificial retina model by two-dimensional multi-layered analog electronic circuits. IEICE Trans. E86-A-2, 387–395 (2003)
9. Kawaguchi, M., Jimbo, T., Umeno, M.: Analog VLSI layout design of advanced image processing for artificial vision model. In: IEEE International Symposium on Industrial Electronics, ISIE2005 Proceeding, vol. 3, pp. 1239–1244 (2005)
10. Kawaguchi, M., Jimbo, T., Umeno, M.: Analog VLSI layout design and the circuit board manufacturing of advanced image processing for artificial vision model. KES2008. Part II, LNAI **5178**, 895–902 (2008)
11. Kawaguchi, M., Jimbo, T., Umeno, M.: Dynamic learning of neural network by analog electronic circuits. In: Intelligent System Symposium, FAN2010, S3-4-3 (2010)
12. Kawaguchi, M., Jimbo, T., Ishii, N.: Analog learning neural network using multiple and sample hold circuits. In: IIAI/ACIS International Symposiums on Innovative E-Service and Information Systems, IEIS 2012, pp. 243–246 (2012)
13. Bengio, Y., Courville, A.C., Vincent, P.: Representation learning: a review and new perspectives. IEEE Trans. Pattern Anal. Mach. Intell. **35**(8), 1798–1828 (2013)
14. Kawaguchi, M., Ishii, N., Umeno, M.: Analog neural circuit with switched capacitor and design of deep learning model. In: 3rd International Conference on Applied Computing and Information Technology and 2nd International Conference on Computational Science and Intelligence, ACIT-CSI, pp. 322–327 (2015)

Study on Category Classification of Conversation Document in Psychological Counseling with Machine Learning

Yasuo Ebara, Yuma Hayashida, Tomoya Uetsuji and Koji Koyamada

Abstract The beginner counselors have difficulty doing to turns interests for the cognitive characteristic and the internal problems by the client, and are using frequency closed-ended question to confirm the interpretation created in ones mind for the client. Therefore, there is the opportunity for education and training which called the supervision to improve the counseling skill of beginner counselor by expert counselors. However, these documents of the verbatim record in the counseling used in the supervision are large-scale and complex, the expert counselors are very difficult to extract the characteristics and situation of the conversation. As appropriate method to visualize each reaction of the client for each question by beginner counselor, we have developed a system for visualizing the flow of conversation in counseling. However, the expert counselor as the system user requires to correct the initial classification result manually, and the work burden is large, because the accuracy of the category classification of conversation document is very low in the current system. To improve this problem, we have implemented on the category classification method for text data of conversation document with SVM (Support Vector Machine) as machine learning technique. In addition, we have compared and evaluated with the result of the initial classification in the current system. As these results, we have shown that the accuracy rate of the classification method with SVM become higher than the result in the current system.

Keywords Psychological counseling · Machine learning · Text classification

Y. Ebara (✉) · K. Koyamada
Academic Center for Computing and Media Studies, Kyoto University, Kyoto, Japan
e-mail: eba@viz.media.kyoto-u.ac.jp

K. Koyamada
e-mail: koyamada@viz.media.kyoto-u.ac.jp

Y. Hayashida
Faculty of Engineering, Kyoto University, Kyoto, Japan
e-mail: hayashida@viz.media.kyoto-u.ac.jp

T. Uetsuji
Graduate School of Engineering, Kyoto University, Kyoto, Japan
e-mail: uetsuji@viz.media.kyoto-u.ac.jp

© Springer International Publishing AG 2018   109
R. Lee (ed.), *Computational Science/Intelligence and Applied Informatics*,
Studies in Computational Intelligence 726, DOI 10.1007/978-3-319-63618-4_9

1 Introduction

In the psychological counseling, the counselor offers counseling with psychosomatic patient or client mainly on the physical symptom from stress causes. Particularly, it is important for counselors to listen carefully while focusing their attention on the interests of the client as the basis skill of counseling [1]. However, beginner counselors have difficulty doing to turns interests for the cognitive characteristic and the internal problems by the client. In addition, beginner counselors tend to proceed with the counseling to suit one's interest, and are using frequency "closed-ended question" (the client can answer "Yes" or "No") to confirm the interpretation created in one's mind for the client.

Therefore, expert counselors need to instruct so that beginner counselors can use an "open-ended question (the question by 5W1H)" freely. In contrast, there is the opportunity for education and training to be called the supervision to instruct for beginner counselor on the counseling skill by expert counselors as the supervisor. In usual the supervision, the supervisor takes a look with direct transcript of verbatim record in the counseling and coaches someone for beginner counselor. However, these documents are large-scale and complex, the supervisor is very difficult to extract the characteristics and situation of the conversation. Accordingly, we considered to be necessary an appropriate method to visualize each reaction of the client for each question by beginner counselor.

To satisfy the request, we have developed the system for visualizing the flow of conversation in counseling [2]. Figure 1 shows an example of the visualization result. This system has visualized the distribution change along the progress of time with the accumulation line graph for each category base on the Adlerian psychology [3] such as "Love", "Work" and "Friend" included in a client's utterances. Each color part in

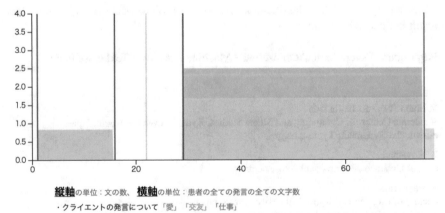

縦軸の単位：文の数、　横軸の単位：患者の全ての発言の全ての文字数
・クライエントの発言について「愛」「交友」「仕事」
・カウンセラーの発言について「開かれた質問」「閉じられた質問」「相づち」「解釈」「世間話」

Fig. 1 Example of visualization result on the flow of conversation in counseling [2]

Fig. 2 Processing procedure
of system for visualizing
flow of conversation in
counseling

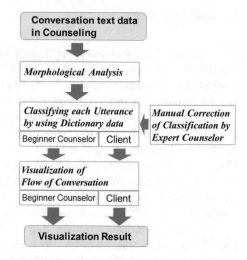

the accumulation line graph is expressed for each category (i.e. pink is "Love", blue
is "Work", and green is "Friend"). In addition, the classification pattern of questions
by a counselor in vertical line have been expressed on the graph. To visualize the
relationship of the utterance between a counselor and a client, the vertical bar graphs
as the classification pattern of questions by counselor are displayed by overlapping
the accumulation line graph. By visualizing the flow of conversation according to
time series in this system, we have shown that it can visually understand how a client
receives the influence of each question by a counselor.

Figure 2 shows the processing procedure on system for visualizing the flow of
conversation in counseling. In this current system, at first, each sentence in text
data which recorded the conversation by a client and a beginner counselor have
been divided for each word by morphological analysis. Then, these sentences are
classified into each category by matching with the keyword corresponding to the
category registered in this system and each word in the sentence.

However, in current category classification method for text data of each utterance
by the counselor and the client, the accuracy of classification results depends on
the word dictionary which registered in system, and the accuracy rate for initial
classification is very low. Therefore, an expert counselor has confirmed the initial
classification result in the current system, and has to correct the wrong classification
results manually, and the work burden is large.

To improve the burden for expert counselor, it is required to automatically classify
the category for text data of the conversation in counseling more accurately. Recently,
the text classification with supervised machine learning technique have been studied
actively. In this paper, we implement on category classification method for text data
of conversation document in counseling with SVM (Support Vector Machine) [4] as
machine learning technique. In addition, we compare and evaluate this classification
results with the initial classification results in the current system.

2   Related Work

The classification of text data is applied for various purposes such as the automatic sorting of spam mails and the automatic classification of news articles. Recently, many studies on the classification of text data using the machine learning technique, and the classification of text data by supervised learning are especially studied.

Taira et al. explained that classification accuracy decreases when not learning considering the words with a low appearance frequency for text classification problem by machine learning, and realized high classification accuracy in the classification of news articles by learning using SVM to use high dimensional word vector [5]. In this study, nouns are extracted from each document of news articles, and documents are vectorized by Bag-of-words model. Bag-of-words is a model that considers only how words are contained in a document without considering the arrangement of words in the document. Previously, Bag-of-words model has been widely used for vectorization of documents with many types of words such as news articles.

However, the text data of conversation document in counseling is classified by each sentence of the utterance, the sentence length is short and number of words is few. Therefore, it is considered that even if each sentence is vectorized using the Bag-of-words model after extracting nouns, it is difficult for the characteristics of the sentence to appear. Moreover, as a disadvantage of the Bag-of-words model, the problem of the specific fluctuation of notation in Japanese and the synonymous words are impossible to recognize as the same meaning is cited.

Mikolov et al. proposed word2vec to vectorize words by learning a distributed representation of words [6]. The word2vec is possible to represent words with high accuracy by closed vector of several hundred dimensions, and various studies on the application are currently being conducted. It is possible to obtain semantic similarity between words by using the distributed representation of words obtained by word2vec, and studies dealing with the meaning of words in Japanese are being conducted. Nozawa et al. proposed a method to discover alternative cooking ingredients by extracting cooking ingredients and cooking methods from a large amount of recipe data, and by calculating words similar to each word from word vectors obtained by learning with word2vec [7]. In addition, Sugawara et al. proposed a method to resolve the synonymy ambiguity of polysemous words using the distributed representation of words [8].

However, there are few studies that create sentence vectors based on word vectors obtained by word2vec and apply them to Japanese sentences classified by machine learning technique. As study cases of document classification dealing with other than Japanese based on the word vector obtained by word2vec, Xing et al. compiled the automatic classification accuracy of Chinese news articles with machine learning technique, after creating document vectors using word vectors obtained by word2vec and word vectors using LDA model respectively [9]. As the result, they showed that the classification method which creates the document vector based on the word vector obtained by word2vec and execute machine learning by SVM was obtained the highest accuracy.

As the classification method in this study, we consider that text data of a clients utterances in counseling are vectorized per a sentence based on the word vector obtained by learning distributed expression of words by word2vec, and automatic category classification is executed by machine learning with SVM.

3 Category Classification of Text Data with Machine Learning Technique

3.1 Overview of Category Classification Method

Figure 3 shows the flow of category classification method with machine learning technique in this study. First, the word vectors are calculated with document data on the troubleshooting of "Yahoo! Chiebukuro" [10] as text corpus by using word2vec [6]. Yahoo! Chiebukuro is a knowledge community and knowledge retrieval service in which participants share with knowledge on electronic boards operated by Yahoo! JAPAN, and about 500 million questions and answers are registered. Based on these word vectors, the supervisor data are generated by vectorizing sentence on the troubleshooting of Yahoo! Chiebukuro with labels of three categories ("Love", "Work" and "Friend"). In the same procedure, the test data is generated by vectorizing the text data per a sentence of the client's utterance in counseling.

The supervisor data is learned as input data of the machine learning, the document data is input to the learned model, and the category classification results is outputted. These results are compared with the modified classification results by an expert counselor on ahead. We explain the details of the classification method in the following Sections.

3.2 Word Vectorization by Word2vec

The data handled in this study is the text data by sentences composed of various words. Most of these data are different in the number of words and the length of sentences. However, input text data is necessary to be expressed as the fixed length numerical vector. In this study, we executed the word vectorization by word2vec which expresses words by the vectors of about several hundred dimensions by learning distributed expressions of words. In general, large scale data such as Wikipedia and news articles in various web sites are often used as the corpus to learn for word2vec. However, the text data of conversation in counseling is usually the conversational sentence, and many adjectives that express clients emotions are included relatively.

Therefore, we considered to obtain large quantities of document data from categories on the troubleshooting of Yahoo! Chiebukuro consultation to use sentences with contents similar to counseling data as the corpus. In particular, we obtained

Fig. 3 Processing flow of category classification of with machine learning

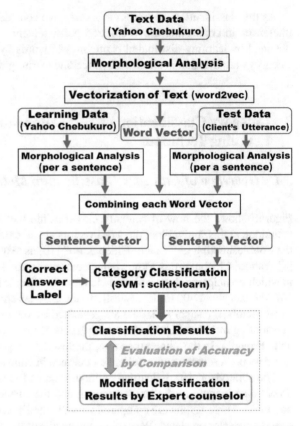

Table 1 Types of text data in troubleshooting of Yahoo! Chiebukuro

Categories	Types of questions
Love	• Consultation for love
	• Distress on family relationship
Friend	• Distress on friend relationship
Work	• Distress in workplace

the four types of document data shown in Table 1 (Each category: 6000, Question sentence: 24,000, Best answer for each question: 24,000) similar in content to the three categories "Love", "Work" and "Friend" classified in this study.

In addition, word2vec is originally considered for English sentences, and each word in the corpus is necessary to be separated by one-byte space. In this study, the sentences of Yahoo! Chiebukuro are separated for a word by using Mecab [11] of the morphological analysis tool.

3.3 Sentence Vectorization

Next, we translate these sentences into numerical vectors using the distributed representation of words calculated from word2vec. The flow of sentence vectorization in this study is shown in Fig. 4. As a procedure of processing, an example sentence in Fig. 4 is categorized by word with morphological analysis and is converted into basic forms. Then, the vector of each word is extracted from the learned model by word2vec for each word appearing in the sentence. The sum of the first dimension of each word vector is calculated, and the sum of all the dimensions are calculated by executing for all dimensions of the word vector.

Xing [9] has shown that the accuracy of classification is improved by a method of creating sentence vectors as an average of simple sums of word vectors obtained by word2vec in the text classification with machine learning technique. Accordingly, the sentence vector is calculated as the average of the sum of word vectors using the same method in this study.

Moreover, the supervisor data is generated by vectorizing for a question sentence with labels of three categories ("Love", "Work" and "Friend"). As document data used for the supervisor data in this study, the question sentences on the categories of "Distress on family relationship", "Distress on friend relationship" and "Distress in workplace" in the troubleshooting of Yahoo! Chiebukuro are separated for a sentence, and 6000 sentences for each category are used.

On the other hand, as document data used for the test data, 170 sentences which classified as the categories of "Love", "Work" and "Friend" per a sentence of the client's utterances in counseling corrected manually by an expert counselor are applied.

3.4 Category Classification with SVM

Next, we explain the category classification method with machine learning technique in this study. As a procedure of classification, first of all, the supervisor data explained in previous Section is learned as an input of the machine learning. Then, the test data in the model after learning is inputted, and one of three categories "Love", "Work" and "Friend" is outputted as the classification result. Moreover, the category classification results which corrected manually by an expert counselor for the initial classification results of the client's utterances in this system are used as the correct label of the test data.

In this study, we used SVM as the algorithm for machine learning. SVM is one of pattern recognition models, and it is a method of configuring pattern classifiers of two categories using the linear threshold elements. Normally, SVM is the binary classification that classifies into two categories. In order to classify into multiple categories in SVM, we applied the One-Vs-The-Rest method. The classifiers to solve

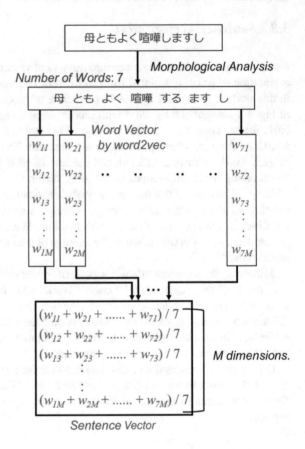

Fig. 4 Flow of sentence vectorization in this study

Sentence Vector

two category classification problems of belonging to a certain category or belonging to one of the other K-1 categories are used in all combinations.

In order to do the learning by SVM, we applied the scikit learn [12] which is an open source machine learning library by python. In this study, the accuracy rate is used as an index to evaluate the accuracy of category classification by this method. The accuracy rate is defined as the ratio of correct category output for all test data. The number of the same output as the correct label in the test data is T, the number different from the correct label is outputted is F. The accuracy rate $Accuracy$ is defined as the following Eq. (1).

$$Accuracy = \frac{T}{T + F} \times 100 \qquad (1)$$

4 Evaluation Experiment

4.1 Setting of Each Parameter in SVM

We conducted the experiment on the category classification using SVM described in the previous Chapter. SVM has different separation hyperplanes by selection of the kernel function and the parameters, and the classification accuracy is also different. Therefore, the accuracy rate of category classification is calculated, when each parameter related to SVM is changed. In this paper, we compare the classification results with SVM by using the parameter with the highest accuracy rate with the initial classification results in the current system. Parameter conditions in this experiment are shown in Table 2.

4.2 Evaluation Results

Table 3 shows the accuracy rate of this category classification for 170 sentences of client's utterances in counseling by changing each parameter condition. From these results, the accuracy rate for 170 sentences become the highest in the condition of the number of vector dimension is 30, $C = 1$ and $\gamma = 10^{-5}$. On the other hand, regarding the number of dimensions, there was no difference in the accuracy rate when the number of dimensions is 30, 40 and 50. As the number of dimensions increases 100, the accuracy rate decreased as compared with the condition where the number of dimensions was low. As for γ, the accuracy rate become high in the case of $\gamma = 10^{-5}$ and 10^{-4} without depending on C. In addition, the accuracy rate become low in the case of $\gamma = 10^{-7}$.

Next, we compared the accuracy rate of this method with the initial classification results in the current system. Each result of the accuracy rate is shown in Table 4. In the result by initial classification, 104 sentences in 170 sentences is unclassified, these unclassified results are classified as uncorrected category. From these results, the accuracy rate of the classification method with SVM in this study is higher than the result in the current system, and we confirmed improvement of the accuracy of

Table 2 Parameter conditions of SVM in this experiment

Parameters	Setting values
Number of dimensions of sentence vector	30, 40, 50, 100
Kernel function	Nonlinear radial basis function
C (Cost parameter)	0.1, 1, 10, 100
γ (Parameter of radial basis function)	10^{-7}, 10^{-6}, 10^{-5}, 10^{-4}, 10^{-3}

Table 3 Results of accuracy rate for each parameter

(a) 30 dimensions

γ					
C	10^{-7} (%)	10^{-6}(%)	10^{-5} (%)	10^{-4} (%)	10^{-3} (%)
0.1	40.0	40.0	48.8	56.5	42.4%
1	38.2	38.2	**63.5**	35.3	39.4
10	38.2	38.2	39.4	37.6	39.4
100	38.2	38.2	39.4	35.3	39.4

(b) 40 dimensions

γ					
C	10^{-7} (%)	10^{-6} (%)	10^{-5} (%)	10^{-4} (%)	10^{-3} (%)
0.1	37.6	37.6	28.8	35.9	41.2
1	37.6	37.6	**60.6**	37.6	38.2
10	40.0	40.0	37.6	38.8	38.8
100	40.0	40.0	40.0	38.8	36.5

(c) 50 dimensions

γ					
C	10^{-7} (%)	10^{-6} (%)	10^{-5} (%)	10^{-4} (%)	10^{-3} (%)
0.1	39.4	39.4	52.4	35.3	40.6
1	36.5	36.5	**59.4**	38.8	38.8
10	37.6	37.6	43.5	39.4	38.8
10	38.8	38.8	38.2	38.8	38.8

(d) 100 dimensions

γ					
C	10^{-7} (%)	10^{-6}(%)	10^{-5} (%)	10^{-4} (%)	10^{-3} (%)
0.1	33.5	33.5	**44.1**	32.4	33.5
1	32.9	32.9	42.9	32.4	34.7
10	31.8	31.8	31.8	34.7	34.1
10	30.6	30.6	35.9	34.1	34.7

classification. Therefore, it is considered that the effectiveness of the classification method with machine learning is shown.

From the results of this experiment, the sentences in which words unique to each category appear alone could be correctly classified by the initial classification in the current system and this classification method with SVM. However, when the word appearing in each sentence is not registered in the dictionary of this system, the initial classification method could not correctly classify. On the other hands, the classification method with SVM in this study could correctly classified these sentences. We consider as the reason why it could be classified that the words included in the sentences of each category of learning data are larger than the learning data of other categories.

Table 4 Comparison results of accuracy rate by both classification methods

	Initial classification in current system	Classification with SVM (30 dim., $C = 1, \gamma = 10^{-5}$)
Number of correct classifications	49/170	108/170
Accuracy rate	28.8%	**63.5%**

However, the accuracy of classification result in this study is insufficient level to reduce the burden of manual correction by the expert counselor as a user. As future work, we will need to further improve the accuracy of category classification.

4.3 Discussion

In this Section, we discuss the cause which could not be correctly classified by the method in this study with examples of characteristic sentences.

Sentence 1–3 shown in Fig. 5 should be the correct classification result as "Love" category, these sentences were classified as "Work" category by the classification method in this study. In this case of Sentence 1, it is considered that sentences including the word "*job*" frequently appear in the learning data of "Work" category, and the word vector of "*job*" may influence on each sentence vector. In order to correctly classify this sentence, we consider that it is necessary to be able to classify into "Love" category by recognizing as "*that person's job*" with the dependency of sentences analysis.

On the other hands, as one of the reasons for the result of Sentence 2, it is considered that the keyword for understanding who is the topic of whom is lacking. In addition, it seems to affect this classification result that there are many sentences in which "*customers*" and "*dishes*" appear together in the learning data of "Work" category.

The word "*design in the future*" in Sentence 3 is divided into "*future*" and "*design*" by morphological analysis. Although the word "*future*" appears more frequently in the learning data of "Love" category, the word "*design*" more often appears in sentences related to the design work in learning data of "Work" category. For example, as there are many sentences such as "*—concerned about my future job*" also in the learning data of "Work" category, it is considered that the sentence "*—can not design in the future*" is often classified as "Work" category. Therefore, we consider that it is necessary to use new dictionary to which the compound nouns such as "*future design*" is added in addition to the default dictionary in MeCab at the morphological analysis.

In Sentence 4, there are cases where category classification is very difficult, because it is difficult to identify who is "*that person*" and other words in the sentence do not exist in learning data of any category. To correctly classify these sentences,

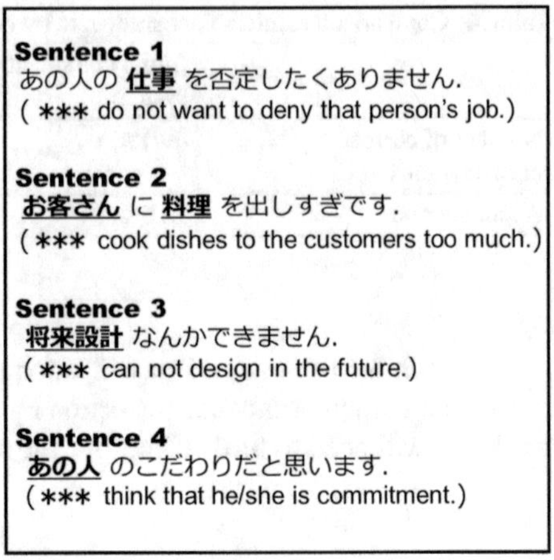

Fig. 5 Example of characteristic sentences that could not be correctly classified

Sentence 1
あの人の **仕事** を否定したくありません.
(∗∗∗ do not want to deny that person's job.)

Sentence 2
お客さん に **料理** を出しすぎです.
(∗∗∗ cook dishes to the customers too much.)

Sentence 3
将来設計 なんかできません.
(∗∗∗ can not design in the future.)

Sentence 4
あの人 のこだわりだと思います.
(∗∗∗ think that he/she is commitment.)

we consider that it is necessary to complement the subject backwards to the previous conversation. In addition, it is necessary to study the classification methods per block of conversation including counselor's questions and previous utterances by the client.

5 Conclusion

In this paper, we have implemented on category classification method for text data of conversation in counseling with SVM as machine learning technique to improve the accuracy of the category classification of conversation document in the system for visualizing the flow of conversation in counseling. In addition, we have compared and evaluated with results of the initial classification method in the current system. As these results, we have shown that the accuracy rate of the classification method with SVM become higher than the result of the initial classification in the current system.

However, the accuracy of classification result in this study is insufficient level to reduce the burden of manual correction by the expert counselor. As future work, we will need to verify the accuracy of classification for more sentences of unknown client's utterances. Moreover, we believe that it is important to propose new classification methods that can correspond to the characteristic sentences that could not be correctly classified by the method in this study.

Acknowledgements We are deeply grateful to the example presenters and clients who had you willingly consent about this example data offer. Special thanks also to the Japan Yoga Therapy Society, for having study support go generously.

References

1. Rogers, C.R.: Client-centered therapy: its current practice, implications and theory. Houghton Mifflin College Div (1951)
2. Uetsuji, T., Imai, S., Onoue, Y., Kamata, M., Ebara, Y., Koyamada, K.: Construction on visualization system of flow of conversation in counseling. In: Proceedings of International Conference on Simulation Technology (JSST2016), pp. 364–369 (2016)
3. Ansbacher H. L., Ansbacher R.: The Individual Psychology of Alfred Adler, Haper Row Publishers Inc, New York (1956)
4. Boser, B.E., Guyon, I.M., Vapnik, V.N.: A training algorithm for optimal margin classifiers, the fifth annual workshop on Computational learning theory, pp. 144–152 (1992)
5. Taira, H., Mukouchi, T., Haruno, M.: Text categorization using suport vector machie (in Japanese), IPSJ Technical Report, 1998-NL-128 (1998)
6. Mikolov, T., Yih, W., Zweig, G.: Linguistic regularities in continuous space word representations, NACCL13, pp. 746–751 (2013)
7. Nozawa, K., Nakaoka, Y., Yamamoto, S., Satoh, T.: Finding method of replaceable ingredients using large mounts of cooking recipes (in Japanese), The Institute of Electronics, Information and Communication Engineers, Technical Report, 114(204), pp. 41–46 (2014)
8. Sugawara, T., Takamura, H., Sasano, R., Okumura, M.: Context Representation with Word Embeddings for WSD, Computational Linguistics. Springer, New York pp. 108–119 (2015)
9. Xing, C., Wang, D., Zhang, X., Liu, C.: Document classification with distributions of word vectors. In: 2014 Annual Summit and Conference on Asia-Pacific Signal and Information Processing Association (2014)
10. Yahoo! Chebukuro. http://chiebukuro.yahoo.co.jp/
11. MeCab: Yet another part-of-speech and morphological analyzer. http://taku910.github.io/mecab/
12. Scikit-learn machine learning in python. http://scikit-learn.org/stable/

Improvement of "Multiple Sightseeing Spot Scheduling System"

Kazuya Murata and Takayuki Fujimoto

Abstract Recently, the number of foreign tourists and travelers coming to Japan is increasing explosively. In 2006, 10 years later, the number of the foreign visitors was approximately 7.3 million people, and it reached approximately 24 million people in 2016. The reason is various Japanese tourism policies such as "Visit Japan Project" and "Cool Japan". As a result, Japan's travel and tourism competitiveness of 2015 was one of the highest in the world. However, a problem is still left in the current Japanese sightseeing situation. For example, there are not many stores with foreign language correspondence. Also there are not many places where they can pay with a credit card. Furthermore, the traffic such as the subway or the buses is complicated. In addition, recently diversification advances for a current Japanese sightseeing method. In the conventional sightseeing, a sightseeing method to visit the famous sightseeing spots listed on the Internet and a guidebook was common. However, particularly these days, as well as a famous sightseeing spots, the number of foreign tourists and travelers to visit the "local-based type" stores such as Japanese bar or Japanese diner, has been increasing intensively. It is not easy to find those kind of places. To solve the problem, we develop "Multiple Sightseeing Spot Scheduling System" that enables a guide of new sightseeing in this research. In this paper, we consider an improvement plan for the application under development and implement the plan.

Keywords Sightseeing · Application · Tourists · Travelers · Sightseeing categories

1 Introduction

Recently, the number of foreign tourists and travelers coming to japan is increasing explosively. In 2006, the number of the foreign visitors was approximately 7.3 million people, and it reached approximately 24 million people in 2016, 10 years later [1].

K. Murata (✉) · T. Fujimoto
Graduate School of Engineering, Toyo University, Tokyo, Japan
e-mail: s46d01300037@toyo.jp

T. Fujimoto
e-mail: fujimoto@toyo.jp

© Springer International Publishing AG 2018   123
R. Lee (ed.), *Computational Science/Intelligence and Applied Informatics*,
Studies in Computational Intelligence 726, DOI 10.1007/978-3-319-63618-4_10

Currently, there are many tourists and travelers coming to Japan for the purpose of "shopping spree". Furthermore, in 2020, "Tokyo Olympics" are going to be held in Japan. From these, the amount of foreign tourists and travelers coming to Japan can be expected to increase in the future. The reason of the increasing number of the foreign visitors is recent Japanese tourism policies. For example, they are "Visit Japan Project", "Cool Japan Policy" and "Relaxation of the visa acquisition for Middle Eastern countries".

The Japan Tourism Agency planned "Visit Japan Project" and Japan National Tourism Organization implements them. "Visit Japan Project" is one of the "strategies to increase the number of foreign tourists visiting Japan". In "Visit Japan Project", Japan established 20 important point markets abroad to attract foreign tourists. Promotions of Japan in those countries or areas are conducted. For example, there are activities such as "informing charms of tourism in Japan by newspapers, magazines and websites" or "inviting the local media to Japan and encouraging them to deliver the attractive points". "Visit Japan Project" also carries out tourism promotion for Japan by various methods besides them.

Cool Japan Policy" is a policy by Ministry of Economy, Trade and Industry. Currently Japan has the famous industries: the car industry, the consumer-electronics industry and etc. In addition, Japan has special cultures including contents such as "Anime", "Manga", fashions and Japanese food. The foreign people appreciates them as so-called "Cool Japan". These kinds of Japanese unique cultures can be expanded to the business deployment properly and it gets people to be interested in Japan more. At the same time, Japan can get foreign demand. Attracting foreign tourists is the activity that can be connected with economic growth of Japan. For Example, for the overseas development promotion of "Anime" and "Manga", there is the localization such as subtitles and dubbing. Also, there is an activity such as the securing of the channel for exclusive use of Japanese contents and the allied product sale in commercial facilities. In this way, the policies target to deliver Japanese various contents to the foreign countries and encourage foreign people to come to Japan.

Regarding "Relaxation of the visa acquisition for Middle Eastern countries", relaxation of the acquisition of the visa is carried out for Middle Eastern countries such as Indonesia, Philippines, Vietnam and China. Particularly, in 2015, the visa acquisition in China was relaxed. By this, a lot of Chinese visited Japan, and "shopping spree," purchasing Japanese products in large quantities was promoted. As a result, "shopping spree" became one of the strong motivation to increase the number of the foreign visitors.

By these Japanese tourism policies, currently, travel and tourism competitiveness of Japan is one of the highest in the world. The sudden increase in the number of foreign tourists and travelers are expected in the future.

2 Purpose of This Research

Currently, travel and tourism competitiveness of Japan is one of the highest in the world. But, a problem is still left in the Japanese sightseeing situation. It is a problem with the information environment on sightseeing spots and stores in Japan.

For example, there are not many places with the foreign language correspondence. Of course, there are a lot of stores with the foreign language correspondence, but the stores with no foreign language correspondence still exist. In addition, there are few places where the credit card payment is possible, although the credit card payment is a basic means of payment in the foreign counties. Furthermore, the traffic such as the subway or the buses in Tokyo is complicated and hard to understand. Regarding sightseeing of Japan, these are the problems.

Furthermore, Diversification advances for a current Japanese sightseeing method. In the conventional sightseeing, a sightseeing to visit the famous sightseeing spots listed on the Internet and a guidebook was common. However, particularly in late years, as well as a famous sightseeing spot, the number of foreign tourists and travelers who visit the sights of the "local-based type" store such as Japanese bar or Japanese diner is greatly increasing. On the other hand, it is not easy to find those kinds of places. Foreign tourists and travelers can check sightseeing information, but elaborate preparation is necessary.

From these, we thought whether we could make the sightseeing for the "local-based type" places easier while solving problems of the sightseeing situation. Based on the thought, we develop "Multiple Sightseeing Spot Scheduling System" that enables a new sightseeing guiding in this research. We explain about the outline of the application and consider an improved plan to implement additional functions.

3 Outline of "Multiple Sightseeing Spot Scheduling System"

In this research, we develop "Multiple Sightseeing Spot Scheduling System" enabling a new tourist guide. In this chapter, we describe the outline of the application of this research.

3.1 Setting for the "Current Place" and "Place to Return"

For the application of this research, unlike common sightseeing application, it is not necessary to set a destination. In this application, as substitute of the destinations, we set "Current Place" and "Place to Return". From the "Current Place" and the "Place to Return", the application determines the scope of the sightseeing places to suggest.

Fig. 1 Setting for "Current
Place" and "Place to Return"

A top screen of this application is indicated in Fig. 1. To set "Current Place", user input the current place in the "Departure" column of the upper part. Similarly, to set "Place to Return", the user input a place to return in the "Arrival" column. In the application, both "Current Place" and "Place to Return" are specified by the name of the stations"

3.2 Setting for "Time of Return"

In the common sightseeing application, the time required to the destination is displayed. In the application of this research the user sets "Time of Return". From "Time of Return" and current time, the system suggests the possible sightseeing routes with calculation. ⌜As a setting method for "Time of Return", the user input "Remaining time" between the current time and time of return in the column of "Time of Return". For example, when the user wants to come back to "Place to Return" two hours later, the user inputs "2" to set "Time of Return" as two hours later (Fig. 2).

3.3 Setting for "Sightseeing Categories"

In the common sightseeing application, users input a destination and search a route. In this application, the user specifies a "Sightseeing category" and decides a sightseeing place without setting a destination. We have set "Sightseeing Categories" such as Table 1.

In this application, with these four components, first, the system "defines the scope from the current place and the place to return". Second, it "calculates time allowance for sightseeing by back calculation from time of return". At last, it "searches appro-

Fig. 2 Setting for "Time of Return"

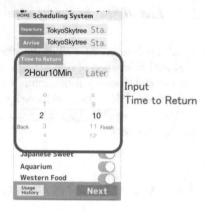

Table 1 Sightseeing Categories

Tourist spot	Cafe	Shrine
Museum	Japanese sweets	History
Aquarium	Japanese food	Souvenir
Archives center	Western food	Garden
Art museum	Chinese food	Park
Memorial hall	Traditional craft	

Fig. 3 Setting for "Sightseeing Categories"

priate sightseeing spots for the specified sightseeing categories". That is how the system suggests possible sightseeing routes in a specified time range.

A list of "Sightseeing Categories" is displayed by at the lower part of the initial screen. The user specifies categories under which he or she wants to see the sights freely form this list. They can specify multiple sightseeing categories (Fig. 3).

3.4 List of "Sightseeing Information"

In this application, as well as the suggestion of the sightseeing route, the user can view the information of sightseeing spots and the stores. In this application, we made a list of sightseeing information to be viewed by the user. In the list of sightseeing information, information on Table 2 is shown.

The users can set sightseeing spots using "sightseeing categories" on this sight-seeing information.

3.5 "Use or Disuse of the Credit Card Payment"

This research aims at suggesting a more comfortable sightseeing route for foreign tourists and travelers. In the application, the users can choose use/disuse of "Credit card payment" when searching a sightseeing route. By this function, we aim to enable the user to enjoy more comfortable sightseeing.

Table 2 List of "Sightseeing Information"

Sightseeing categories	Store name	Address	TEL	Opening hours
Regular closing day	Required time	Required money	Credit card	

Fig. 4 "Use or disuse of the credit card payment"

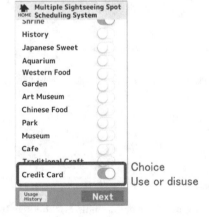

3.6 Search Results

When the user has set five components: "current place", "place to return", "time of return", "sightseeing categories", and "use or disuse of the credit card payment", and carried out a search, the screen image in Fig. 4 will be displayed. At first, the search results of the sightseeing routes by the text are displayed. When the user chooses a favorite sightseeing route from the list and taps the arrow, application will start the guide of the sightseeing route on the map. Then, the user can view information by tapping the name of a sightseeing spot or the store. The user can choose a favorite sightseeing route in consultation with this information.

The guide of the sightseeing route on the map is indicated in Fig. 5. The user can enjoy sightseeing during a specified time while checking this map.

The user can view information if the user taps the name of the stores on this screen (Fig. 6).

4 Improvement Plan for "Multiple Sightseeing Spot Scheduling System

In the series of this research, we previously carried out a simple comparison experiment for the three sightseeing methods: "Sightseeing using a guidebook"; "Sightseeing using a smartphone"; "Sightseeing using the application of this research." We also reviewed three sightseeing methods and investigated improvement of the application.

Fig. 5 Search result by the text

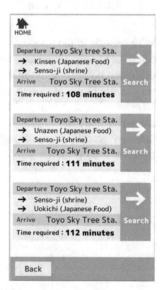

Fig. 6 Search result on the
map

In this paper, regarding "Multiple Sightseeing Spot Scheduling System" under development, we consider an improvement plan from the results of the comparison experiment. We also implement the improvement plan to the application. In this chapter, we consider functions to be improved in the application of the current developmental stage.

4.1 Addition of the "Reviews" Function

With regards to the application, when choosing sightseeing spots or stores, the user can refer to only homepages and sightseeing information created in this research. He or she can obtain only basic information. As a result, it is still the case that the user has very limited information to take into account. To solve the problem, we considered "Reviews" as an additional function plan. Using this "Reviews" function, the user can write feely the impressions and evaluations of the sightseeing spots and stores that they actually went for sightseeing. Other users can choose sightseeing spots or stores of their preference more easily to refer to the comments of the "Reviews".

4.2   Addition of the "Priority Ranking" of the Sightseeing Categories

In the application of this research, a user is required to choose sightseeing categories freely and the system suggests the sightseeing plans based on the search for the sightseeing spots or stores that belong to the categories of the user's choice.

In the present application, even if a user chooses multiple sightseeing categories, the sightseeing spots or stores that belongs to those categories are randomly listed in the sightseeing plan. To solve this, as an improvement plan, we add a function of "Priority Ranking" to sightseeing categories.

By a function of "Priority Ranking", for example, there is the situation: "First, I want to have lunch of Japanese food. Then, I want to see the sights of the Shrine". In this case, the user needs to choose "Japanese food" under the sightseeing category first. Next, the user needs to choose "Shrine" under the sightseeing category. By this function, the user can create the sightseeing plan of their preference more accurately.

4.3   Addition of the "Remaining Time Alarm" Function

In the present application, the time during which a user can see the sights is limited by a function of "Time of Return". The user needs to see the sights while thinking about the remaining time by himself or herself. To improve this, as an additional function, we implement "Remaining time alarm". For example, in the case of the remaining time at 30 min, it is a function to promote attention awakening such as "Remaining time is 30 min" as an alarm function. By this function, the user can see the sights more comfortably with conscious awareness of time left.

5   Implementation of the Improvement Plan for "Multiple Sightseeing Spot Scheduling System"

In this chapter, we describe functions implemented according to the improvement plan of the application that we mentioned earlier in this paper.

5.1   Addition of the "Reviews" Function

Regarding the "Reviews" function, we added "Reviews" column to the screen that only simple sightseeing information was shown previously. In reference to those, the user can choose sightseeing spots and stores.

Fig. 7 Sightseeing
information

As a viewing method of "Reviews", "Reviews" contributed by other users is listed on the lower part of the screen of the sightseeing information.

Next, we explain about how to contribute to "Reviews". The user can contribute to "Reviews" from the usage history of the top screen. In a usage history, the history of the sightseeing route that a user used in the past is saved. From here, the user can contribute to "Reviews" about the sightseeing spots and stores. "Reviews" that a user contributed to will be reflected to the page of the sightseeing information as shown in Fig. 7.

By tapping "Reviews" button on the page of the usage history, the contribution screen will be displayed (Fig. 8).

Next, the user needs to tap the button of the sightseeing spot or store on the contribution screen and then, a contribution form will be displayed. The user can contribute the impression or evaluation to "Reviews" using this contribution form (Fig. 9).

5.2 Addition of the "Priority Ranking" of the Sightseeing Categories

Addition of "Priority Ranking" of the sightseeing categories changed the system specification.

In the previous version of the application, we adopted the way to choose sightseeing categories using switch functions, By the improvement, we abolished switches

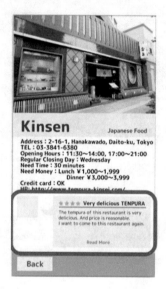

Fig. 8 Example of "Reviews"

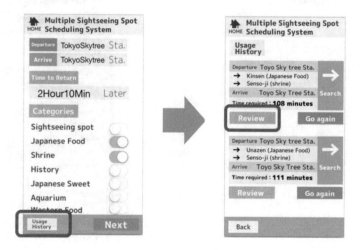

Fig. 9 Contribution method for "Reviews" 1

and changed the system to enable the user to choose sightseeing categories by picker views (Fig. 10).

To set the priority, the system requires the user to choose sightseeing categories in the priority order from the top. For example, there is a user's situation: "First, I want to have lunch of Japanese food. Then, I want to see the sights of the Shrine". In this case, the user chooses "Japanese Food" in "First Category" column and chooses "Shrine" in "Second Category" column. By this function, the application can provide a sightseeing plan that reflects the user's preference in better way (Fig 11).

Fig. 10 Contribution method for "Reviews" 2

Fig. 11 Improvement of "Sightseeing Categories"

5.3 Addition of the "Remaining Time Alarm" Function

In the former simple evaluation experiment, the remaining time of the sightseeing was described in the search results. The user had to do sightseeing while thinking about time to see the sights by own. As a result, the user was able to do sightseeing

Fig. 12 "Remaining time alarm" function

smoothly. However, it is not the case that all of the users can control time completely. To improve this, we implemented the alarm for the remaining time. By this alarm function, the user can know the remaining time automatically and anyone can see the sights more smoothly within the time limit (Fig. 12).

6 Conclusion and Future Work

In this research, we are developing "Multiple Sightseeing Spot Scheduling System" enabling guidance of new form of sightseeing. In this paper, from the results of a previous simple evaluation experiment, we considered an improvement plan for the application. Based on the plan, we implemented three systems. By these systems, the user can see the sights more comfortably within limited time left.

The future development of this research has a large-scale evaluation experiment. In this paper, we implemented improved systems, but have not yet carried out a large-scale evaluation experiment. As the next step, we will carry out a large-scare evaluation experiment using the application. From its results, we want to investigate further improvements for the problems of the system and implement them.

Acknowledgements This work was supported by JSPS KAKENHI Grant Number 17K00730

References

1. Japan National Tourism Organization: Trend of the number of the visit to Japan foreign visitors. http://www.jnto.go.jp/jpn/reference/tourism_data/visitor_trends/
2. The World Economic Forum: The Travel & Tourism Competitiveness Report 2015. http://www3.weforum.org/docs/TT15/WEF_Global_Travel&Tourism_Report_2015.pdf
3. Japan Tourism Agency: Inbound Travel Promotion Project (Visit Japan Project). http://www.mlit.go.jp/kankocho/en/shisaku/kokusai/vjc.html
4. Ministry of Economy, Trade and Industry: Cool Japan/Creative Industries Policy. http://www.meti.go.jp/english/policy/mono_info_service/creative_industries/creative_industries.html
5. Ogawa, K., Sugimoto, Y., Naito, K., Hishida, T., Mizuno, T.: Basic design of a sightseeing recommendation system using characteristic words. IPSJ SIG Tech. Rep. MBL **71**(14), 1–6 (2014)
6. Urata, M., Nagao, S., Kato, F., Endo, M., Yasuda, T.: Photo rally system to support tourists in tourism areas. J. Jpn. Inf. Cult. Soc. 21(2), 11–18 (2014)
7. Mizukami, T., Hayashi, T., Ikari, S., Hishida, T., Mizuno, T.: Implementation of Sightseeing Application COMAT base on Agricultural Information. IPSJ SIG Tech. Rep. MBL **70**(48), 1–8 (2014)
8. Mizukami, T., Hayashi, T., Sugimoto, Y.: COMAT: citizens cooperation mapping for toyota. IPSJ SIG Tech. Rep. MBL **69**(2), 1–4 (2013)
9. Jalan: Introduction of the tour guide application-Jalan net. http://www.jalan.net/jalan/doc/howto/iphone_kankou.html
10. Kuroda, S.: TABIMARU Tokyo. Shobunsya Publications, Inc., August 2015
11. K. Murata, T. Fujimoto.: Proposal of multiple travel scheduling system based on inverse operation method. In: 14th IEEE/ACIS International Conference on Computer and Information Science 2015 (ICIS 2015), pp. 503–507, June 2015
12. Murata, K., Fujimoto, T.: Development of "multiple sightseeing spots scheduling system" and comparison with the existing sightseeing methods. In: The 14th International Conference on Software Engineering Research and Practice (SERP 2016), pp. 230–235, July 2016

Advertising in the Webtoon of Cosmetics Brand-Focusing on 'tn' Youth Cosmetics Brands

Sieun Jeong, Hae-Kyung Chung and Cheong-Ghil Kim

Abstract Recently in order to establish a brand image of a cosmetics company and to communicate emotional communication with adolescents, advertisements using Webtoon are increasing. Therefore, this study suggests ways to improve the effects of advertisement setting and story structure, character setting and story structure which can attract youth's interest by studying advertisement analysis of youth cosmetics brand Webtoon. This study focuses on a small number of female students attending Beauty high school and it is worth the studying because of the opportunity to look closely at their thoughts and requirements of the Webtoon advertisements through Focus Group Interview method with young people.

Keywords Cosmetic brand · Webtoon advertisement · Youth cosmetic

1 Introduction

Recently, the breakthrough in social media and social networking has brought about the phenomenon of diversification and individualization, and the interest in beauty is increasing. As a result, the information about makeup that has been limited to adult women spreads to young people and gradually became established [1]. According to the statistics released in the 2016 Korean Journal of Hairdressing, 93.4% of teenage girls suffer from basic cosmetics and more than half of teenage girls actually use lip gloss or lipstick. Although the school officially regulates the makeup of students, it is our reality that students are increasingly interested in and consuming cosmetics.

S. Jeong
Department of Design, Konkuk University, Seoul, Republic of Korea
e-mail: sieunjeong@naver.com

H.-K. Chung
Department of Moving Image Design, Konkuk University, Seoul, Republic of Korea
e-mail: jangmi44@gmail.com

C.-G. Kim (✉)
Department of Computer Science, Namseoul University, Cheonan, Republic of Korea
e-mail: cgkim@nsu.ac.kr

© Springer International Publishing AG 2018

R. Lee (ed.), *Computational Science/Intelligence and Applied Informatics*,
Studies in Computational Intelligence 726, DOI 10.1007/978-3-319-63618-4_11

Such a phenomenon can be seen as a kind of fashion-like culture beyond making ornaments and beauty of youth themselves. The interest of young people who are organized as cosmetics leads to the purchase of cosmetics and plastic surgery. It increases the interest in the cosmetics market and the students who become more competitive due to the child-centered purchase behavior due to the nuclear family become the subject of final payment by themselves without their parents' advice. This means that the youth group occupies an important position as a consumer of cosmetics.

Today, cosmetics brand companies that have teenagers as main customers are interested in marketing methods for promoting the purchase of young people in accordance with the changes and trends of the times. Therefore, the domestic cosmetics industry is showing exclusive products for various teenagers. Cosmetics companies targeting teenagers are concentrating on emotional marketing using 'Webtoon' that can satisfy various sensibilities and needs of girls rather than using star marketing or distribution network. As a marketing method for a cosmetics brand company, 'Webtoon' has attracted attention with a variety of storytelling structures, but it is actually inferior to the expectation of high school girls. As this interest grows, some researches were conducted to investigate the relationship between the purchase intention of the adolescents and the purchase behavior of the adolescents. However, there is very little qualitative research on marketing methods for young people. Therefore, it is necessary to study the direction of advertising using the 'Webtoon' in order to establish a brand image of a cosmetics company and to communicate emotional communication with adolescents.

This study was selected as a type of classical drama advertisement + creative storytelling according to the branding type of 'Webtoon' in terms of storytelling advertisement of Yoon (2015). Therefore, we will select 'Teenager HoonReo Club' of 'tn', a specialized brand of cosmetics for teenagers, as a classical drama ad + creative storytelling type, and analyze contents and form. The content analysis includes images, characters, conflicts, and plots. The analysis of forms deals with advertisers, advertising purposes, and advertising strategies. Based on the analysis, we constructed a questionnaire and conducted a qualitative research method focus group interview. We tried to explore the meanings of the experiences of the actual users of the 'Webtoon' and the experience of the branded 'Webtoon'. Through these researches, youth cosmetics brand can be a way to enhance character setting and story structure that can attract interest and experience from adolescents'.

2 Contributions of Current Study

2.1 Webtoon and Branded Webtoon

Webtoon is a coined word meaning 'cartoon made on the web', and which is combined with 'web' and 'cartoon'. With the use of the Internet, Webtoon has become more active for consumers based on smartphones. The number of companies that use this

Table 1 Branded webtoon

	Innisfree	Innisfree	Tonimori	Gudal
Webtoon title	'Speak with your lips'	The 'Coach of the Poisonous Water Forest'	'The world is too dry for women'	'The real moisturizing story'
Story	The story of 20-year-old female college campus romance begins with the choice of lip color	The heroine 'Kwon' is struggling to save the goddess of water in a barren world	It can be solved through essence of skin anxiety which every woman is suffering from	Create a variety of stories on the subject of 'Essence', a feature of oil essence
Webtoon image				
Webtoon writer	Park So-hee	Kim Cheol-hyun	Kim Jin	Himam Ali
Story genre	Romance	Warrior story	Common	Common

Webtoon as a means of marketing advertising is becoming increasingly popular. In particular, cosmetics companies of small and medium-sized companies emerge from the application of cartoon and characters to products, and empathize consumers with various story settings and familiar contents that can lead to customer's emotions. The brand Webtoon should actively carry out the purpose of publicity of the work itself, and should make consumers' interest and curiosity about the brand together with the desire to purchase [2].

In the marketing method using the Webtoons, there is a brand Webtoon that has a one-cut banner advertisement at the end of the serial Webtoon and advertises the brand and the storytelling characteristic with the intention and the story of the brand. Most of the branded Webtoons have publicity, but most of them are indirectly presented by the campaigned Webtoons, which are naturally presented in the story [3]. However, the branded Webtoon can cause adverse effect to the readers due to the excessive publicity of the setting, which may affect the damage of the brand image. Therefore, the cosmetics company should be careful to adjust the degree of publicity of the branded product to be implied and the setting of the brand Webtoon story together with the story writer and the picture artist. As the Table 1, Innisfree is showing genuine, martial art genre of webtoons, and Tony Mori and Gudal have developed a story that can be enjoyed by teenagers in everyday life.

2.2 Storytelling Advertising

Story telling has a characteristic of stimulating emotions by indirectly presenting the characteristics, function, quality, and merits of products or brands to consumers through stories [4]. This may allow consumers to pay attention on the company or product, and they become interesting and fun. Companies are also helped to improve their brand image and have differentiated from other companies.

2.2.1 The Types of Storytelling Advertising

The types of storytelling advertising are classified into 'classical advertising drama' and 'vignette advertising drama' by Stern (1994) [5]. According to Park, Myung-Jin (2012), classical drama commercials are based on the traditional stories that are familiar to us, such as the drama, causal composition, and conflict factors of characters [5]. And it is said that the vignette drama advertisement is presented with various characters, time, place, etc., and the relationship in the character is unclear, and it is seen as an advertisement through various acts [6]. In storytelling of advertising, it is important to select a variety of stories to represent the contents of the brand or product. A brand targeting teens needs access to information about the product and a story that satisfies the trends of the current generation [8] (Table 2).

2.2.2 Analysis of Cosmetics Brand 'tn' Webtoon

The researchers selected 'tn' among the various cosmetics brands displaying the Webtoon. The reason for this is that 'tn' is the cosmetics that teenagers use, and that only does it have two or more Webtoons that are published on a portal site that is accessible to teenagers. And also it is a brand that can naturally realize consensus with students due to using subject in school e pisode [11].

Table 2 The content type of storytelling advertisement [7]

1. Real storytelling	(1) Consumer experience
	(2) True story overcoming adversity
	(3) The myth created by CEO
2. Parody storytelling	(1) Literature
	(2) Movies
	(3) TV programs
	(4) Art objects
3. Creative storytelling	(1) Story that might be in reality
	(2) Exaggeration based on reality
	(3) The world of imagination

Teen's Nature is a cosmetics brand, 'tn', for young people created by Yuhan-Kimberly. 'Fongdang HoonReo Club' and 'Teenager HoonReo Club', are branded Webtoons of 'tn', serialized in NAVER's Webtoon service of Korean main portal site, and stories about youths as cosmetic motifs in high school. In this study, we analyzed the Webtoons of the 'Teenager HoonReo Club' [12].

In Table 3, it is about the contents analysis of 'tn' Webtoon 'Teenager HoonReo Club'.

The writings and pictures of the 'Teenager HoonReo Club' are composed of 8 pieces in total by writer skill, and the genre of the Webtoon is the episode form that includes the fantasy elements in the daily life of the school. The morphological analysis is the Webtoon corresponding to the classical drama advertisement, and the narrative has a causal composition that proceeds according to the passage of time. Classical drama advertising shows the process of solving the conflict between the characters through the acting of the branded model. In 'tn' Webtoon, the main character, Yoo Cholok, meets Tienyunni and becomes a causal link to 'tn' Cosmetics so she did no thicker makeup and got good healthy skin [13].

3 Research Method

The participants of this study were Focus Group Interviews on 18 students from 2 grade 1, 10 grade 2, and 6 grade 3 students attending a beauty high school in Seoul. For focus group interviews, Morgan (1996) defined "as a research method for collecting data on research topics set by researchers through group interaction." Krueger (1986) defined focus group interviews as "Organized group discussion" [9]. As a qualitative method, Beck, Trombetta, and Share (1986) stated that "an informal discussion between selected people related to a particular topic" [14].

Focus group interviews on the opinions of readers about the Webtoons made in the brand position are rarely conducted. We selected focus group interviews for high school students, who are the main customers of youth brand cosmetics brand Webtoon advertisements, so that they can explore their thoughts within a wide range of research purposes. Therefore, this study is meaningful work by collecting and analyzing diverse opinions of interested readers interested in cosmetic brand Webtoon.

3.1 User Test

In Table 4, the students who participated in the study consisted of 18 female students attending S beauty club and J beauty club. In the case of the first research subjects, the ratio of the number of general high school students and beauty high school students was adjusted according to the ratio of the number of students. However, general high school students said that they subscribed to Webtoons as a general characteristic, but they did not interested in cosmetics brand Webtoons. The researchers tried to

Table 3 Content analysis of 'tn' webtoon 'Teenager HoonReo Club'

Advertiser		Cosmetics brand 'tn'
Advertising purpose		It aims at understanding and understanding of importance of basic care in 10 makeups
Title/Author		Teenager HoonReo Club/Giryang
Series duration and number of times		February 19, 2014–April 9, 2014 (8 times)
Type		Classical drama advertisement + creative storytelling
Advertising strategy		The creative storytelling material that could be in reality raised the intimacy of students. In addition, the main character Yoo Gyulk exposed tn cosmetics to realize the importance of basic care makeup, and the process of finding the beauty of real skin was solved by causal narrative structure. Through this process, we are enabling young readers to recognize the tn slogan 'basic skin, teen skin'
Webtoon content	Image	
Analyze content by component	Character	Yoo Cholok: This is a student who has transferred to beauty private school as heroine. She enjoys the usual make-up and wants to know the secret of the beauty of Tienee Leader. She met leader Tienee and receive cosmetics 'tn', and realize the importance of basic care Tienee: As a leader of the hunter's club, deliver tn cosmetics to the oeuvre, which overlooks the importance of basic care. It helps us to understand the importance of basic care Public Health Center Teacher: It is the person who informs about the secret of the hunting club. You will find out about this female club through your health center teacher Banya: Beauty characterization As a popular boy student of private college, only the tn foundation product of Yu-gil comes up against the right figure Students in the same class: Hearing the abstracts that have received Teeny's attention and hiding the colored cosmetic pouch bags. In the end, this occasion will realize that Yu Yeulok is beautiful without having to make up his own hue
	Conflict	Conflicts have been triggered by the students in the same class hiding the cosmetic pouches that are important to the abstract. In the end, only the basic cosmetics of tn are made right. In this process, the banyas are inferior, and in the abstract, the conflict is resolved by discovering their beauty through Tien
	Plot	Yoo cholok transferred to the beauty private school, and this story started. And the story ends with the transfer to the private school, and the content ends as I realize that even basic care alone can be beautiful
	Message	Basic care through cosmetics is important

analyze the students after subscribing to Webtoons for general high school students, but they excluded students from general high school during the research because they were not interested in analyzing cosmetics brand Webtoon because they were not interested.

The participants selected 3 h of makeup theory and practice lessons per week and 4 h or more of skin theory and practical instruction in regular classes. Webtoons subscriptions were composed of research participants who showed interest in Webtoons more than 7 days a week, subscribed more than 20 min a day. In addition, the age composition of the study participants is 2 in the first grade, 10 in the second grade, and 6 in the third grade. Kim, Sung Jae (2000) proposed 12 out of 6 participants without appropriate competition among the participants 11. The focus group was composed of six people. The researcher explained the purpose, content and purpose of the research to the subjects who expressed their intention to participate in the research. The research subject and contents were explained and confirmed once again through the e-mail to the subjects who expressed their intention to participate in the research [15].

3.2 Data Collection Process and Research Procedure

Based on the research of Yoon, Nara(2015) [16], the researcher classified the 'Teenager HoonReo Club' of 'tn' as a type of classical drama advertisement and creative storytelling. Data collection period was 3 months from September 7, 2015 to December 10, 2015, and the data collection was collected through the focus interview between the target groups.

3.2.1 Research Procedure

(1) Preparation
   The researcher arrived at the meeting place 30 min before the start of the interview, arranged the desk in a circle, and confirmed the location of the tape recorder. In addition, after screening of the scrolling screen of the '10 Daughters' Club screening, participants were asked to share their interview questionnaires and be fully informed.
(2) Purpose of study introduction
   The researcher proceeded as an interviewer and explained that the information derived from the interview process is used only as a resource for research. I asked for your understanding of recording and explained the purpose of the interview in detail.
(3) Notes during interview
   During the interview, if the participants' remarks were persistent, the topic of the debate was reminded once more and the research flow was not cut off.
(4) Summary of interview contents

Table 4 General characteristics of participants

Number	1	2	3	4	5	6	7	8	9	10	11	12	13	14	15	16	17	18
Grade	1	1	2	2	2	2	2	2	3	3	3	3	2	2	2	2	3	3
Age	17	17	17	18	18	18	19	18	19	19	19	19	18	18	18	18	19	19
Webtoon subscriptions	2 more times	4 more times	1 more times	10 more times	2 more times	5 more times	10 more times	4 more times	10 more times	5 more times	4 more times	10 more times	5 more times	10 more times	4 more times	4 more times	1 more times	10 more times

After discussing the topic, we briefly summarized the discussion and confirmed it to the participants and proceeded to the next topic. After all the interviews, I thanked the participants for the discussion.

(5) Data transfer

As the researchers proceeded with the interviews, the responses and the atmosphere of the research participants who responded to the interviews were gathered together with notes. The interviews were conducted immediately after the group interview, and each warrior work was completed within five days of the interview. The results of the interviews that have already been done have been revised the types of questions with reference to the following interviews. The researchers participated in the focus group, listened carefully and recorded the process, and clarified the ambiguity with the following questions. We had three processes of transferring all the data obtained during the interview process and reading out the data and identifying important information. The identified information was categorized by conceptualizing them by theme.

(6) Analysis of data

This study was analyzed by Krueger (1998). Before the focus group, which was the first step of the research, the researchers analyzed the branded Webtoons through the previous research and related books to determine the main and sub questions. In the second stage, we participated in the focus group directly. In order to secure the validity of the data, the ambiguity was revised through the questioning process again. In the third stage, immediately after the focus group, the contents of all the processes during the discussion were checked, and additional data were collected through individual phone calls or e-mails about the shortcomings [17].

In the fourth stage after the end, the recording was transferred immediately, and the contents of the discussion were repeatedly heard. Through the four steps, the researcher coded the categorization according to the question in the interview, and the central meaning was the topic coding. In order to confirm the credibility of the interview, the applicant interviewed the two doctoral students of the Advertising Design Graduate School and one of the current Webtoon writers. In addition, the contents analysis was set up with one person who was learned in the analysis of qualitative research [18].

(7) Coding work

After reading the collected data again, I made a list based on the central theme that repeatedly reads the participants' thoughts, such as dialogue, behavior pattern, and thinking style of the participants. In this process, coding categories were found and these coding categories became a good way to classify the data [19]. The researchers were able to find major topics according to the coding categories and organize them by serializing them under the topics. At this time, a list of subheadings was created by repeating the process of adding and integrating subjects by adding numbers related to new words, types, and behaviors. Table 5 shows the results obtained through this iterative process.

Table 5 List of categories

Major subject	Minor subject	Meaning
Story of webtoon (41)	1. Story material (24)	1. Excessive settings (16), Situations that are not appropriate for student status (14)
	2. Word selection (33)	2. Does not match with content (28), Uses outdated swallow (12)
Webtoon characters (28)	1. Realism (33)	1. There is a distance from high school students (21)
	2. Drawing (21)	2. Requires the need for detailed character description (18)
Brand advertising strategy (33)	1. Overreaching (41)	1. Unnatural public relations (22), Public relations (8)
	2. Product description (29)	2. Lack of product information (26)

4 Results

4.1 Topic 1 Story of Webtoon

Participants enjoyed Webtoons of Naver and Daum portal site and answered that they use dedicated app as well as Internet. The paintings and dialogues on the 'tn' brand's Webtoon were complained that it was too much of a setup and a story that was far from their own. In addition, they pointed out the difficulty in the content of the story that the words used in the contents do not use the terms used by the young people and the contents are not understood well.

I see Webtoon very often, usually when I drop out, I look in the subway. By the way, 'tn' Webtoon is not comfortable. This is not our target; it's like a cartoon for the children.

I think it's too much to change the main character into beautiful girl after using 'tn' Cosmetics. (Participant 2)

I do not use these words like 'club jjang'. The words are not the same as the words we use. (Participant 10)

I have a hard time setting the desk as a vanity. I think it would have been much better if the story was developed in a situation that could really happen in high school. (Participant 4)

It is funny that all the students except YooCholok are carrying 'tn' basic cosmetics in their pouches. 'tn' basic cosmetics brand also should be a branded Webtoon that tells about the actual cosmetics like 'makeup story for the 'Daesaleo'. (Participant 12)

4.2   Topic 2 Characters of Webtoon

Most of the interviewees said that the character of the 'Teenager HoonReo Club' is not realistic. The researchers showed the phenomenon of identifying themselves with the protagonist and gave a negative opinion about the perfect appearance of the character.

I do not understand why Yoocholok is carrying cat ears. I like the way the picture is, but the character is too beautiful, so I don't feel realistic. Just like a cartoon. (Participant 1)

The main character Yoocholok is a pretty. She must be pretty whether color makeup or a foundation. I also want to have the same eyes. (Participant 3)

It does not make sense that Teeny looks like a high school student. If you are wearing like Teeny's short skirts and makeups in school, then you should definitely be punished in school. (Participant 6)

The heroine eyes and clothes makeup are too simple. I think it would be nice if you could draw this part in realistic detail. (Participant 17)

4.3   Topic 3. Brand's Advertising Strategy

The researchers pointed out that the distinction between branded webtoons to advertise 'tn' brand and promotional webtoons for product promotion is not well distinguished. In addition, the participants of the study said that the promotion of the products on the webtoons story is unnatural, and that the 'tn' products and the webtoon itself are neglected.

I do not want to be naturally promoting in the content of the webtoon, but since the product is so leaning up, I have lost my desire to see the webtoon anymore. Is this a 'tn' brand webtoon? Or is it product promotion? I cannot tell which is. (Participant 8)

I do not think that I have a desire to buy 'tn' cosmetics even if I look at webtoons from 1st to 8th because I cannot see how much moisturizing and changing the skin.

I do not think it's like bibby cream, but I did basic care. I think I'm overpriced for the product. (Participant 11)

The researchers focused on the result that the protagonist was beautiful after using the basic care within the contents of the webtoon, and he was sorry that he did not provide detailed information on how to use the product or the product.

Yoocholok became beautiful after she is the make-up of color or only basic care? I think the contents are finished too hastily ... There is no explanation about the product of 'tn' cosmetics. I would have liked to have added some special beauty tips like 'GetitBeauty'. (Participant 10)

Skin care like emulsion, etc. When to apply, season, skin type ... Add something like this to help you choose a basic product (Participant 12)

The basic care products of 'tn' cosmetics have only moisture effect? It seems to emphasize only the moisture here. I can expect a variety of effects on basic products.

For example, naturalistic cosmetics, if you have fun with the use of good materials in the webtoon, will not it increase the quality of the 'tn' product? (Participant 11)

In our design approach, we are concerned with all issues that go into providing an engaging and enjoyable experience for people in both short and longer term. This includes aesthetics, pleasure, and emotional engagement [10].

5 Conclusion

WebToon has a unique image and novel story that makes it easier for consumers to feel sympathy and has the advantage of being able to watch anywhere and anytime without restriction of time and place. It also provides its customers with new and memorable differentiated value. It is effective in strengthening of brand [20]. The conclusions and suggestions derived from the data analysis so far are as follows. The main character Yoo Cholok was exposed to 'tn' cosmetics and realized the importance of basic make-up makeup. The process of finding the beauty of real skin was solved by causal narrative structure, but it was suggested that it is difficult for students to sympathize with the story. In the process of subscribing to the branded Webtoons, high school students' use of over-established settings or older generation terms not used among adolescents proved to be an impediment to empathy with the contents of the Webtoon. Therefore, it is necessary to develop and construct a Webtoon that includes the culture of youths to construct branded Webtoons. In character analysis, there is a lack of reality. The students did not agree that the characters had a perfectly beautiful appearance, and they knew that they wanted a character whose projection was present. Therefore, the youth cosmetics branded Webtoon is required to set realistic characters that do not fall short of the environment, condition and distress of the students.

Visual information has increasingly been used to enable human-human communication and knowledge discovery with the vast information. Visual information communication and interaction synergises state-of-theart research in visual communications, designs and applications. By marrying multi-disciplinary research works in visualization, graphical user-interface, and interaction together with art concepts and designs, it has opened a new opportunity to present and analyse information on different perspectives [13].

In addition, the negative opinion that the advertisement about the Webtoons of the cosmetics brand is unnatural and the communication is not done and the antipathy occurs. Although the purpose of advertising is to increase awareness and understanding of the importance of basic care in the ten major makeups, consumer students wanted specialized knowledge on ingredients and usage of cosmetics. I would like to try using it. The negative aspects of the content of the product that do not meet the expectation of the youth and cause the antagonism because the youth do not provide the desired elements on the webtoon are considered to be improved in the future. This study focused on a small number of female students attending Beauty Private high school. However, because of the opportunity to look closely at the thoughts and requirements of their web-to-do advertisements through in-depth interviews with young people, it is worth the study. It is expected that the follow - up studies will be carried out together with the quantitative and qualitative researches that extend the study subjects and target areas in the future.

Acknowledgements We wish to express great appreciation to Konkuk University for their help in share this results and findings. "This paper was supported by Konkuk University"

References

1. Moon, A.J.: Determinants of the turnover intention by hair shop environment and development plan, pp. 184–189. Konkuk University Master's Thesis (2016)
2. Nerurkar, O.: Designing sustainable fashion: role of psychosocial factors of fashion consumption and the challenges of design. Indian J. Sci. Technol. **9**(15), 1–7 (2016)
3. Park, M.J.: The influence of advertising storytelling types on emotional involvement, pp. 3–27. Ph.D. Thesis, Korea University (2012)
4. Yoon, A.: Study on the types of branded webtoons in terms of storytelling, pp. 21–37. Hongik University Master's Thesis (2015)
5. Stern, P.N.: Eroding grounded theory. In: Morse, J. (ed.) Critical Issues in Qualitative Research Methods, pp. 34–39. Sage, Thousand Oaks, CA (1994)
6. Krueger, R.A.: Developing Questions for Focus Groups. Sage, Thousand Oaks, CA (1998)
7. Beck, L.C., Trombetta, W.L., Share, S.: Using focus group sessions before decisions are made. North Carotina Med. J. **47**, 44–50 (1986). Ti-A
8. Yun, S.L.: Adolescents 'buying patterns of cosmetics: focusing on the effects of cosmetics advertising model on high school girls' purchase of cosmetics, pp. 70–72. Dongduk Women's University Master's Thesis (2012)
9. Bapu, B.R.T., Gowd, L.C.S.: Security over the wireless sensor network and node authentication using ECCDSA. Indian J. Sci. Technol. **9**(39), pp. 24-33 (2016)
10. Alireza, R., Jared, D.: Design of a tangible data visualization. Int. J. Softw. Inform. **9**(1), 51–59 (2015)
11. Raam, K.V.J., Rajkumar, K.: A novel approach using parallel ant colony optimization algorithm for detecting routing path based on cluster head in wireless sensor network. Indian J. Sci. Technol. **8**(16), 1–7 (2015)
12. Jin.: Characteristics of TV commercial drama: type, expression, narrative. Research on Advertising Research, vol. 18, no. 1, pp. 40–45 (2007)
13. Quang, V.N., Yingcai, W., Weidong, H., Tomasz, B.: Visual information communication and interaction. Int. J. Softw. Inform. **9**(1), 1–2 (2015)
14. Hiroki, N., Shota, S., Teruhisa, H.: Recognition and intensity estimation of facial expression using ensemble classifiers. Int. J. Networked Distrib. Comput. **4**(4), 203–211 (2016)
15. Sook-hee, Y., Sun-nam, C.: A study on the therapeutic course experience of youth art therapists-using focus group interviews. Art Ther. Res. **22**(2), 20–22 (2015)
16. Kim, S.J., Kim, H.J., Lee, K.J., Lee, S.: Focus group research method, Hyunmosa, pp. 31–35 (2000)
17. Bogdan, R.C., Biklen, S.K.: Qualitative Research in Education: An Introduction to Theory and Methods, 3rd edn. Allyn & Bacon, Needham Heights, MA (2010)
18. David, L.M.: Source: Annual Review of Sociology, vol. 22, pp. 129-152 (1996)
19. Dongsong, Z., Anil, J., Lina, Z., Isil, Y.: Context-aware multimedia content adaptation for mobile web. Int. J. Networked Distrib. Comput. **3**(1), 1–10 (2015)
20. http://www.the-pr.co.kr/news/articleView.html?idxno=10542. (2016.05.19.)

Testing Driven Development of Mobile Applications Using Automatic Bug Management Systems

Mechelle Grace Zaragoza, Haeng-Kon Kim, In-Han Bae and Jong-Hak Lee

Abstract Software development testing practices contain a little pragmatic evidence to support the utility of test-driven development that has been circulating for years. As testing mobile applications servs as a process by which application software developed for mobile devices are tested for its usability, functionality especially consistency. These applications require test automation for the reason of compatibility and speed. Mobile applications either come pre-installed or can be installed from mobile software distribution platforms. Increasing complexity of the mobile applications system makes difficult to test and evaluate the quality properly. Resulting to the automated testing methodology that is becoming popular, resulting to a becoming outdated of manual testing. Model-Driven Testing Techniques artifacts software engineering bases on the model transformation principle. This implies increasing research on automation of the testing processes. In this paper, we proposed an approach to derive tests from the model of the mobile applications system as well as the a diagram in using automatic bug management system. Using this technique, we can achieve more effective testing on hardware related software areas.

Keywords Testing driven development · Mobile applications · Automatic bug management

M.G. Zaragoza · H.-K. Kim (✉) · I.-H. Bae · J.-H. Lee
Catholic University of Daegu, Gyeongsan, South Korea
e-mail: hangkon@cu.ac.kr

M.G. Zaragoza
e-mail: mechellezaragoza@gmail.com

I.-H. Bae
e-mail: ihbae@cu.ac.kr

J.-H. Lee
e-mail: jlee11@cu.ac.kr

© Springer International Publishing AG 2018   151
R. Lee (ed.), *Computational Science/Intelligence and Applied Informatics*,
Studies in Computational Intelligence 726, DOI 10.1007/978-3-319-63618-4_12

1 Introduction

Mobile applications Testing is the most important factor in its software development. Mobile applications testing is a process by which application software developed for hand held mobile devices is tested for its functionality, usability and consistency. Mobile applications either come pre-installed or can be installed from mobile software distribution platforms. Increasing complexity of the mobile applications system makes difficult to test and evaluate the quality properly. As a result, automated testing methodology is becoming popular and in turn decline of manual testing. Because of the characteristics of Mobile applications software, automated testing has difficulty performing all relevant tests and evaluation of the areas of concern. Model-Driven Testing Techniques (MDT) artefacts software engineering bases on model transformation principle. This implies increasing research on automation of the testing processes (Fig. 1).

Evolution and repair program are the main components of today's maintenance software, which consumes a daunting fraction of the total cost of software production. Automated techniques to reduce costs are therefore beneficial for this type of approach. Huge software developers must con-firm, order and limit bugs or bug fixes first and validate fixes [1]. A bug in software terminology is an error, a flaw, mistake or a failure in a computer program that produces ambiguous or unexpected results, or produces unplanned mode. Most bugs and mistakes made by people in the source code of a program or its design [2]. There is no perfect software because it means the software may require an additional module or the existing form or module to upgrade may contain an unrecognized error that remains in the software in its process. The

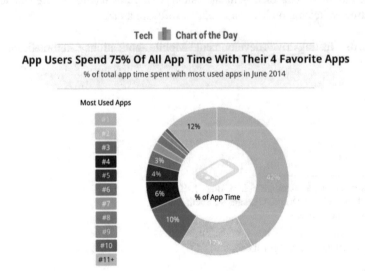

Fig. 1 App usage time spent

bug can be introduced at all stages of software development, and this is something that software developers should consider looking after [3].

Dated July, new data from Nielsen said there's an "upper limit" to how many apps a person will use each month—roughly 22 to 28 apps, on average. Reports suggest that's when people use their phone, 75% of the time they are using one of their four favorite applications. A consumer's most-used application is used about 42% of the time—more on of a messaging application like iMessage, WhatsApp, or Facebook. Users' other top three most-used apps are used 17, 10, and 6% of the time. Users generally spend like 12% of their phone time with all of their other lesser-used apps [4].

In this paper, we present an approach to derive tests from the model of mobile applications system. The analysis of methodologies used for mobile applications as well as for standard system development demands creation of a bridge between them. We also will discuss the reliable testing processes. In particular, test development for each phase of system engineering is proposed. Input signals as continuous, discrete and real time constraints are the factors indicating object oriented or function oriented approach. Finally, Model Driven Testing ideas are mentioned so as to elaborate the full overview on test process automation for Mobile applications systems.

2 Related Works

2.1 Bugs Life Cycle

There are many bug tracking systems available in the industry for use. Monitoring systems are also called bug monitoring system issues or system failure reporting or system error detection or defect notification system, etc. Bug tracking systems have been developed by an open source community and organizations like proprietary closed software [3].

2.2 Mobile Applications Software Bug Management Process

Mobile applications software testing is a disciplined process that consists of evaluating the application (including its components) behavior, performance, and robustness. One of the main criteria, although usually implicit, is to be as defect-free as possible. Expected behavior, performance, and robustness should therefore be both formally described and measurable. Verification and Validation (V&V) activities focus on both the quality of the software product and of the engineering process. These V&V activities can be sub-classified as preventative, detective, or corrective measures of quality. While testing is most often regarded as a detective measure of quality, it is closely related to corrective measures such as debugging. In practice,

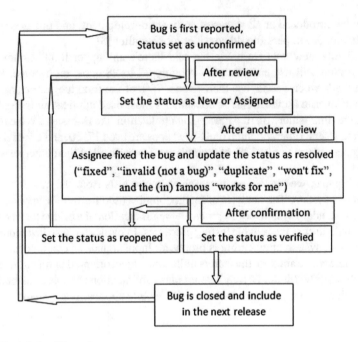

Fig. 2 Generic bug life cycle

software developers usually find it more productive to enact testing and debugging together, usually as an interactive process. Bug management literally means removing defects (Fig. 2).

Mobile applications software industries in airborne, train, and automotive domains is becoming more mature, and catch up with other industries such as the computer hardware industry. Model based development processes are established and exercised for real-time Mobile applications systems. Nowadays, there is a strong trend to automate the safety-critical functions, which in turn requires application of safety standards. Automation demands the application of formal methods and formal verification. That is why the development process of Mobile applications systems considers the question which process steps in the product life cycle have to be covered [5] and supported by appropriate tools. The most important steps are:

- **capturing** of textual specification
- **modeling** of software and hardware topology and its functions considering all the constraints automatic code generation from the models automatic test code generation from the test models
- efficient software development, **testing and debugging** environment Mobile applications software engineers and test engineers have to develop and verify their software using model checking together.

The corresponding benefits cope with design failures, which are detected early in the overall process. The quality of specification models increases. These improvements result in significant cost reductions during the software development process. The Mobile applications software design methods used for years suffered from informal specifications, lack of adequate support for verification, fairly long design times. This situation has become untenable as their complexity and safety, cost and power consumption requirements put on them has scaled up. The situation has been made even more difficult by the increasing degree of integration in the semiconductor industry that has made possible to build Systems-on-Chips (SOC) with unparalleled compute power. In too many cases, errors in conception and implementation of Mobile applications controllers have caused dramatic problems especially in the area of space exploration and applications. On the other hand, the opportunities offered by technology for Mobile applications controllers are immense.

2.3 Problem of Manual Handling

Bug reports consist of activities such as surveying, gathering information, testing, and debugging throughout the process. It is very difficult to manually manage the problems of a project, because hundreds of bugs can be found. Developers create bugs, but the QA team is one that examines the code and the application to find out the exact sequence or combination of steps that generate an error.

2.4 Ineffective Communication

All development teams need effective communication systems. It is impossible for anyone to keep all bugs on top or in a single document, such as spreadsheets. And they cannot communicate effectively with each other or with the development team and therefore do not contribute to increasing the quality of the product. As the project develops, the first problem that may occur is that a person can edit the sheet at a time. Communication is a vital and decisive factor for the success of a software project.

2.5 Complex Systems

Nowadays, a large number of similar bug tracking systems available for this purpose; and most of them are created with the complex and unnecessary confusing features. Flow is not simple is a primary concern. Most bug tracking systems are built into a web application. Web-based applications can often be difficult to navigate, especially if they behave more like an HTML page sequence than an application. Some web applications constantly recharge pages as the user input response requires a round trip between the web server and the user's browser. This leads to slow down the page refilling process and makes the application difficult to use [6].

3 Mobile Applications Bug Management

The trend is clear: there is an urgent need for automatic techniques to complete manual software development with inexpensive tools. Research into automated repair programs focused on reducing repair costs by allowing the program to continue with run-time errors, but [7] the problem with the old system can be defined as project maintenance, user maintenance, and assignment must be kept manually. Software development companies are faced with many problems while manually maintaining all maintenance projects, their errors, and their status.

This kind of problem makes the entire system ineffective and creates poor and unorganized work. To eliminate this problem, the document is expected to be upgraded. Bug tracking software is a"bug tracking system" or a set of scripts that keep a problem reporting database. Bug Tracking Software allows individuals or groups of developers to effectively monitor outstanding product bugs. Bug Tracking Software can track bugs and changes, communicate with members, submit and review patches, and manage quality assurance. This commercially-based web application is a great tool for assigning and tracking issues and tasks during software development and other projects involving teams of two or more [5].

3.1 Mobile Applications Debugging and Testing Management (MADT)

Mobile Application Debugging and Testing (MADT) is a disciplined process of evaluating the behavior, performance, and robustness of the application (including its components). One of the main criteria, though usually implied, should be free from defects as possible.

The Fig. 3 shows the system architecture of mobile Applications debugging and testing development. It consists of repository and test case generation for mobile source codes. The source code is analyzed in the host and these test cases are generated. Test cases are then sent to a target table using various methods and finally executed. Test case results are then sent to the host and analyzed as Fig. 4.

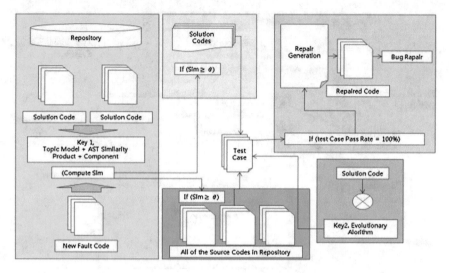

Fig. 3 System architecture of MADT (Mobile Application Debugging and Testing)

Fig. 4 System structure of
MADT execution
environment

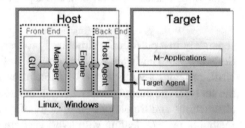

3.2   Model Driven Testing for Mobile Applications Software

In usual software engineering approaches, object oriented modeling with UML has
become popular as in Fig. 5. In this area, object orientation is extremely advantageous:
oriented object modeling to structure application data so as to facilitate maintenance
and spread the responsibility of some functions between participating objects or
components. Finally, object-oriented techniques can use a modern design model that
improves flexibility and durability. In the field of mobile application systems such
as Simulink models [8] or languages as a short-term hardware description language
at very high speeds (also VHSIC hardware description language), VHDL [8] is used
to describe hardware circuits. C is used to program electronic control units (ECUs).
Programmable logic controllers are programmed into a kind of assembler, using
functional block diagrams or Pascal as structured text.

They usually involve some kind of pointer concept and dynamic memory heap
organization. Pointers may be zero or indicate memory cells already paid. Consumption can increase uncontrollably. Many functional requirements in a car electronic
control unit (ECU) cannot be handled at the same time. Thus, these aspects are treated

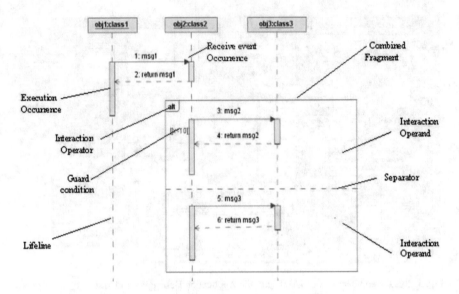

Fig. 5 UML architecture to use for mobile applications software testing tool

separately in a profoundly structured process, eg. User Functions, Communication, Software and Hardware Architecture as well as component software distribution. UML provides a wider range of description media, with particular advantages for the analysis and design phases. In Simulink, they are needed to achieve the necessary abstractions and topology descriptions of clearly defined guidelines and modeling methodology. By dealing with hybrid (continuous and discrete) signals, with additional real-time conditions, it is difficult to choose the methodology / technology for designing a mobile application system model. In addition, real-time constraints on the system side are still valid for the trial model. That is why decision-making for system development affects the test. In the case of standard software, these effects are not so critical because the test methodology is independent of the system development methodology. In the case of real-time mobile application software, we take care of the simulation based on the system model. This simulation is defined on the side of the system, but from a certain point of view it hits the test. For example, model simulation runs provide template cover information that helps test groups determine which aspects of implementations are covered by equivalent evidence. Test data and vectors can be generated for use in test harness, either directly from needs or by design. This is especially important for systems that contain a lot of logic, where design of test sequences is particularly difficult.

3.3 From Specification, Through Models, Towards Bugging and Testing

Debug and testing specification can be partly derived from system specification. This implies formalization of the system requirements. Such requirements-driven testing enables test teams to develop tests against current requirements, rather than building them in isolation. Since testing is based on conformance to requirements instead of general test statistics, this approach delivers higher quality results: test teams can validate that the system, software or product does what is required. The approach is provided by DOORS tool such as the IBM Rational DOORS which is a is a multi-platform, enterprise-wide requirements management tool, designed to capture, link, trace, analyze, and manage a wide range of diverse textual and graphical information to ensure a project's compliance to specified requirements and standards [9]. The automated transfer of information appears, however manual methods to retrieve the final test requirements are used.

MADT gives the possibility to design and implement functions of time continuous behaviour, while discrete behaviour or modeling the architecture on abstract level can be done in UML (but also in Simulink). Depending on the system engineers one has to adapt testing methodology to their requirements. One of the options is to use UML 2.0 models as completion of MDT models for Mobile applications systems. Both kinds of models should be mapped and developed parallel completing each other and taking into account system requirements as given in Fig. 6. Model transformation between the different philosophies of object orientation in UML and functional orientation in Simulink is still a matter of research. In this paper, we promise practical solution in mid-term future. Simulink model is tested during its simulation. It can be additionally tested using Classification Tree Editor for Mobile applications Systems (CTE/ES) and Mobile testing tools. UML 2.0 models

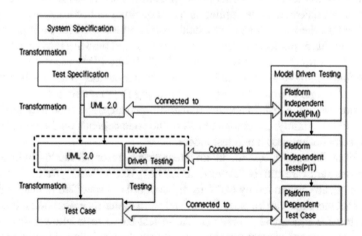

Fig. 6 Mapping between debugging and testing specification and MADT

(e.g. HybridUML Profile) must be additionally tested with traditional approach. The testing methodology against UML models is the application of UML 2.0 Testing Profile (U2TP) being the Object Management Group (OMG) standard. Test artefacts retrieved from UML models and applied by UML 2.0 Testing Profile models serve as the base for further testing procedures. Moreover, the interfaces between Simulink models and UML models should be also tested, at least partly automatically. All the connections found, have to be tested. Further, at least two ways are possible— derivation of test code from the test model or derivation of test code from the system implementation code. The first one corresponds to Model Driven Testing concepts, that is why only this will be investigated. It is again a challenge, as the information must be derived not only from the UML 2.0 Testing Profile test models, but also from Simulink model and simulation. This can be done by using the formal methods. All the testing processes described till now correspond to black box testing. That is why the good candidate for test implementation code applied for real-time systems is extended version of Test and Testing Control Notation version 3 (TTCN-3). Finally, implementation code generated either manually or automatically from the models must be tested in white box testing. Although in the second case it has been proven that the code provides fewer failures than in the first case (Fig. 7).

MADT on Model Driven Development prescribes certain model artefacts used along system development line, how those models may be prepared and their relationship. It is an approach to system development that separates the specification of functionality from the specification of the implementation of that functionality on a specific technology platform. Main MDA artefacts are platform independent system models (PIMs), platform specific system models (PSMs) and system code. There is a clear distinction between PIM, PSM and system code although it depends on the context, the development process and the details of the system and target platform, where the border between PIM, PSM and system code is to be placed. Within these three abstraction levels, transformation techniques are used to translate model parts of one abstraction level into model parts on another abstraction level. These MDA abstraction levels can also be applied to test modelling as according to the philosophy of MDA, the same modelling mechanism can be reused for multiple targets. As shown in Fig. 5, platform independent system design models (PIM) can be transformed into platform specific test models (PIT). While PIMs focus on describing the pure functioning of a system independently from potential platforms that may be used to realize and execute the system, the relating PITs contain the corresponding information about the test. In another transformation step, test code may be derived from the PIT. Certainly, the completeness of the code depends on the completeness of the system design model and test model.

The following code is Test conductor integration tests for interactions with Mobile applications hardware and model with MDT. The "_Expect" and "_Return" functions are automatically generated by MDT from the interfaces specified in header files.

Test specifications can be partially derived from system specifications. It is about formalizing the system requirements. These test-based tests allow test teams to develop tests against current requirements rather than building them in isolation. Since tests are based on compliance with requirements in place of general test

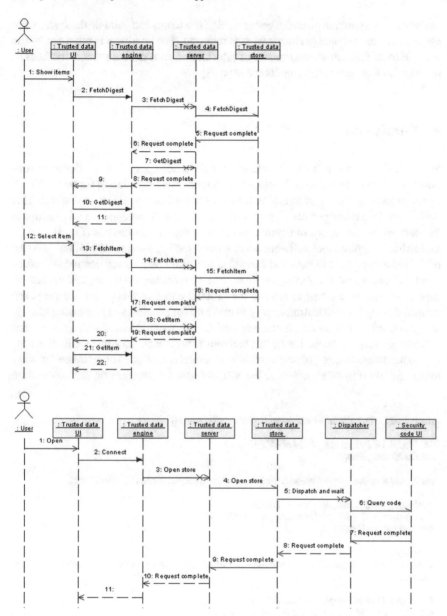

Fig. 7 Sequence diagram for MADT on Model Driven

statistics, this approach provides better results: test teams can validate the system, the software, or the product performs what is required. The approach is provided by the PORTE tool. The automated transfer of information appears, but manual methods to recover the final test requirements are used [8].

4 Conclusion

Software development testing practices contain a little pragmatic evidence to support the utility of test-driven development that has been circulating for years. These applications require test automation for the reason of compatibility and speed. This Automatic Bug Management Systems will ensure that these apps are all relevant to the devices and the operating system versions in the market there is (Fig. 8).

In Mobile applications software development, testing and quality evaluation is one of the most important factors and can affect the entire Mobile applications development process. These factors require lot of time, so reducing the testing and evaluating time is an effective factor to release the product early. In this paper, we proposed manual debugging and testing as way to make up for problems of automated testing. We proposed model based on existing problem of manual based testing areas and analysis the requirements. Using this technique, we can achieve more effective testing on hardware related software areas. Test results can be used as resource for other testers for sharing of experience. We also proposed a diagram on the automation

```
static void testRunShouldNotDoAnythingIfItIsNotTime(void)
{
AdcModel_DoGetSample_Return(FALSE);
AdcConductor_Run();
}
static void testRunShouldNotPassAdcResultToModelIfSampleIsNotComplete(void)
{
AdcModel_DoGetSample_Return(TRUE);
AdcHardware_GetSampleComplete_Return(FALSE);
AdcConductor_Run();
}
static void
testRunShouldGetLatestSampleFromAdcAndPassItToModelAndStartNewConversionWhenItIs
Time(void)
{
AdcModel_DoGetSample_Return(TRUE);
AdcHardware_GetSampleComplete_Return(TRUE);
AdcHardware_GetSample_Return(MDTU);
AdcModel_ProcessInput_Expect(MDTU);
AdcHardware_StartConversion_Expect();
AdcConductor_Run();
}
```

Fig. 8 Proposed testing driven development of mobile applications using automatic bug management systems diagram

of debugging and testing development for mobile applications. MADT on Model Driven Testing ideas are used so as to elaborate the full overview on testing automation process intended for mobile applications systems. Future work requires further investigation on test case information retrieval. This implies a lot of work in the context of transformation, mapping rules and technical possibilities, test implementation code and identifying bugs for time continuous behavior is still in the process of development. Research on this will allow us to investigate the relativity between the object and function-oriented design on system model and the test model side. In Mobile applications software development, testing and quality evaluation is one of the most important factors and can affect the entire Mobile applications software development process. These factors need a lot of time, so reducing the testing and evaluating time is basically an effective factor to release the product as early as possible and is minimal or bug free.

Acknowledgements This Research was partially supported by Catholic University of Daegu. Korea 2017 Foundation.

References

1. Le Goues, C.: A systematic study of automated program repair: fixing 55 out of 105 bugs for $8 each. In: 2012, 34th International Conference on Software Engineering (ICSE). IEEE (2012)
2. Muhammad, Y.J., Hufsa M.: An automated approach for software bug classification. In: Complex, Intelligent and Software Intensive Systems (CISIS), pp. 414–419 (2012)
3. Singh, V.B., Chaturvedi, K.K.: Bug tracking and reliability assessment system (btras). Int. J. Softw. Eng. Appl. **5**(4), 1–14 (2011)
4. Myungmuk, K., et al.: Improvement of software reliability estimation accuracy with consideration of failure removal effort. Int. J. Netw. Distrib. Comput. **1**(1), 25–36 (2013)
5. Smith, D.: CHART OF THE DAY: Most People Use Only 4 Apps, Business Tech Insider, Sep. 5, (2014)
6. Saifan, A.A., et al.: Test case reduction using data mining technique. Int. J. Softw. Innov. (IJSI) **4**(4), 56–70 (2016)
7. Zhang, D., et al.: Context-aware multimedia content adaptation for mobile web. Int. J. Netw. Distrib. Comput. (2014)
8. Yamada, A., Mizuno, O.: Classification of bug injected and fixed changes using a text discriminator. Int. J. Softw. Innov. (IJSI) **3**(1), 50–62 (2015)
9. IBM https://kr.mathworks.com/products/connections/product_detail/product_35525.html
10. Sandeep, S.: Analysis of bug tracking tools. Int. J. Sci. Eng. Res. **4**(7), 134 (2013). ISSN 2229-5518
11. Le Goues, C., Forrest, S., Weimer, W.: Current challenges in automatic software repair. Softw. Qual. J. **21**(3), 421–443 (2013)
12. Fiaz, A.S., Devi, N., Aarthi, S.: Bug tracking and reporting system (2013). arXiv:1309.1232
13. Kim, H.K.: Test Driven mobile applications development. In: Proceedings of the World Congress on Engineering and Computer Science 2013, WCECS 2013, pp. 23–25, San Francisco, USA, 2 (2013)
14. Haeng-Kon, K., et al.: Effective mobile applications testing strategies. Int. J. Future Gener. Commun. Netw. **9**(11), 317–326 (2016)

Shape Recovery of Polyp from Endoscope Image Using Blood Vessel Information

Yuji Iwahori, Tomoya Suda, Kenji Funahashi, Hiroyasu Usami,
Aili Wang, M.K. Bhuyan and Kunio Kasugai

Abstract Endoscope is used to remove the polyp in the medical diagnosis. Absolute size of polyp has been usually estimated by medical doctor with their empirical judgement using endoscope. However this estimation depends on the experience and skill of medical doctor and it is sometimes necessary to use the medical thread with known size for estimating the size of polyp. This paper aims to help medical doctor by proposing a new approach to estimate the size and 3D shape of polyp as a medical supporting system. This proposed approach uses blood vessel as a target with a known size to estimate the absolute size of polyp. Using sequential two images make it possible to estimate the movement of endoscope and reflectance parameter. The idea of using blood vessel is the key idea of this paper, where color information,

Y. Iwahori (✉) · H. Usami
Department of Computer Science, Chubu University, 1200 Matsumoto-cho,
Kasugai 487-8501, Japan
e-mail: iwahori@cs.chubu.ac.jp

H. Usami
e-mail: usami@cvl.cs.chubu.ac.jp

T. Suda · K. Funahashi
Department of Computer Science, Nagoya Institute of Technology, Gokiso-cho, Showa-ku,
Nagoya 466-8555, Japan
e-mail: suda_g@cvl.cs.chubub.ac.jp

K. Funahashi
e-mail: kenji@nitech.ac.jp

A. Wang
Department of Communication Engineering, Harbin University of Science and Technology,
Harbin, China
e-mail: aili925@hrbust.edu.cn

M.K. Bhuyan
Department of Electronics & Electrical Engineering IIT Guwahati, Guwahati 781039, India
e-mail: mkb@iitg.ernet.in

K. Kasugai
Department of Gastroenterology, Aichi Medical University, Nagakute, Aichi 480-1195, Japan
e-mail: kuku3487@aichi-med-u.ac.jp

© Springer International Publishing AG 2018
R. Lee (ed.), *Computational Science/Intelligence and Applied Informatics*,
Studies in Computational Intelligence 726, DOI 10.1007/978-3-319-63618-4_13

labeling, morphology processing are used estimate the size and 3D shape of polyp as a final goal. Experiments with endoscope images are demonstrated to evaluate the validity of proposed approach.

Keywords Svm · Multiple classes · Defect candidate region · Defect classification

1 Introduction

In recent years, detection and recovery of polyp from endoscope image has been developed in the research field of computer vision. Especially 3D shape is important as the key to judge the status of polyp and it is necessary to estimate the correct size and shape of polyp from endoscope image. To estimate the size of the polyp present in the endoscope image, skill and experience of medical doctor are mainly affected to the medical diagnosis without computer assisted system. Especially misjudgement of estimating size of polyp sometimes becomes a serious problem since removing method of a polyp depends on its size.

Shape from Shading (SFS) is an important approach and it uses the image intensity as a key to recover the shape of a target object from a single image. Horn [3] pioneered the development of SFS in computer vision, then many approaches have been proposed.

As a recent approach to estimate the size and shape of polyp in endoscope image using SFS, paper [5] has been proposed. This paper proposes a method to recover the shape of polyp from two endoscope images to estimate unknown reflectance parameter after converting the original color images into Lambertian images. The method modifies the surface gradient from the relative shape and tries to recover the absolute size and shape of polyp.

Alternatively, another extension is proposed by [12], where the reflectance parameter is treated as a known constant. These papers introduces optimization approach to estimate surface gradient parameters (p, q) using both photometric and geometric constraints of Lambertian reflectance. Z is determined by (p, q) towards the neighboring points after some initial points with local maximum intensity are selected as starting points.

This paper further extends these approach. The goal is to recover the absolute size and shape of polyp under the condition that reflectance factor is unknown. Instead of treating the amount of movement of endoscope as known constant, the proposes approach uses information of blood vessel observed in endoscope image to estimate the amount of movement of endoscope in Z direction, which results in estimating the reflectance parameter C. The approach makes it possible to recover the absolute size of polyp as a result. Two images are used to realize this purpose. The approach extracts blood vessel candidate and it is evaluated by computer experiments using endoscope images.

This paper extends [10] from the point of view that reflectance parameter is unknown and blood vessel candidate is selected to estimate the amount of movement of endoscope. and finally the reflectance parameter is obtained. Given this parameter, the shape is recovered by the optimization approach [12].

We have introduced GrowCut [6] to detect blood vessels in the previous research [10]. GrowCut is one of the segmentation approaches using a cellular automation. GrowCut needs the surrounding region to be specified as blood vessel by user. This means it takes time and costs user's manual operation although the manual operation to surround the blood vessel region is satisfactory. Here this paper proposes an approach to detect blood vessel effectively and tries to recover the absolute size and shape using the information of detected blood vessel. The assumption is that target object has smooth surface with Lambertian reflectance converted from the original endoscope image [4, 9].

1.1   Detection of Candidate of Blood Vessel

Detection of blood vessel consists of two step processing. The first step detects the blood vessel by the color information and labeling in the endoscope image. Procedures for detection of blood vessel is shown in Fig. 1.

Fig. 1 First stage of vessel detection

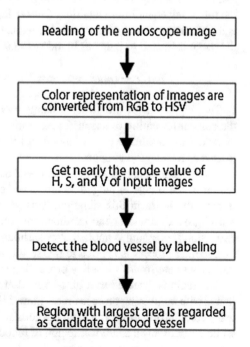

Fig. 2 Result of blood vessel detection using color information

0	0	0	1	0	0	1	0
1	0	0	0	0	0	0	0
0	0	0	0	0	0	0	1
0	0	1	0	0	0	1	1
0	1	1	1	0	0	0	1
1	1	1	0	0	0	0	1
0	1	0	0	0	0	0	0
0	0	0	0	0	0	0	1

Two endoscope images are used. These two images are taken under the same environment and the second image is that with the small movement of the first image along the depth direction. The basic condition required is that polyp and blood vessels are observed in these two images.

Color representation of two images are converted from RGB to HSV. Hue, Saturation and Value are extracted from the converted first image with scanning. The criteria for which pixels are judged as blood vessel is given to the pixels which has nearly the mode value of each of H,S and V.

Not only one pixel but other pixels which satisfy the following range are actually extracted. The range(Ves) which is judged as the blood vessels is a pixel that satisfies the following equation. In addition, $Center$ is each mode of HSV respectively, max is the maximum value of HSV, and the parameter α, which represents how much of the range is treated to judge as blood vessels, is used as known.

$$Center - (max \times \alpha) <= Ves <= Center + (max \times \alpha)$$

Two input images are scanned using Ves obtained by the above equations. If the color information of scanned pixel satisfies the range of Ves, then the pixel is extracted as a candidate pixel of blood vessel, otherwise as a background. Here black is used as a background color.

In the image of candidate of blood vessel detected by color information, too small region in the image is judged as blood vessel. Shrinking processing is applied to remove the noise in this situation. This processing can treat too small region as a background, however the candidate region of blood vessel become also small, expansion processing is applied after shrinking processing.

Labeling process is applied to this result image. There are several labeling techniques but here Imura's labeling processing [8] is used this paper.

The regions detected as a blood vessel are colored by the labeling in the input image with numbering on each area. Figure 2 shows a result of blood vessel detection using color information in an input image as an example. 0 represents the background, while 1 represents a candidate region of blood vessel.

Fig. 3 Result of labeling
and opening processing

0	0	0	0	0	0	0	0
0	0	0	0	0	0	0	0
0	0	0	0	0	0	0	2
0	0	1	0	0	0	2	2
1	1	1	1	0	0	0	2
1	1	1	0	0	0	0	2
0	1	0	0	0	0	0	2
0	0	0	0	0	0	0	0

Labeling process is applied and candidate of blood vessel is detected to the image of Fig. 2. It is confirmed that the continuous region and non-continuous region exist in Fig. 2. Both region is judged as blood vessel candidate but small region may affect worse the refinement process of blood vessel. Therefore, this small region is removed at this stage. This removing is done by expansion and shrinking processes. The larger region where only the blood vessel is remained and labelling process is applied with numbering to each region. Figure 3 shows the result of labeling processing.

From the area of each region where numbering is applied, region with largest area is regarded as candidate of blood vessel and other region is regarded as background. There are two blood vessel regions in the Fig. 3. Here, the region with number 2 has the smaller area than that with number 1 and region with number 2 is regarded as background. As a condition of numbering the candidate of the blood vessel, the region with one pixel up, down, left and right of the pixel as a blood vessel candidate is excluded.

Candidate of blood vessel in the second input image is detected in the same way as that in the first image. To detect the candidate region of blood vessel in the second image, search is applied from the near the final candidate region of blood vessel in the first image, Search range is taken as the region inside the rectangle region which covers whole candidate region of blood vessel in the first image. Even if the multiple candidates of blood vessel exist, only the largest area is extracted as a candidate. If there is no candidate inside the region, rectangle region is expanded. Expanding ratio is taken with β percentage expansion at a time to each of x and y coordinate of four points which constructs the rectangle. Coordinates of four points which constructs square are obtained as follows. Example is shown in Fig. 4. Point A to D indicates the coordinates of rectangle with yellow color while A' to D' indicates the coordinates of rectangle with blue color.

$$A = (x_{min}, y_{min})$$
$$B = (x_{max}, y_{min})$$
$$C = (x_{max}, y_{max})$$
$$D = (x_{min}, y_{max})$$

Fig. 4 Example of
rectangular region

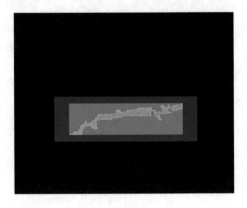

$$A' = (x_{min} \times (1 - \beta), y_{min} \times (1 - \beta))$$
$$B' = (x_{max} \times (1 + \beta), y_{min} \times (1 - \beta))$$
$$C' = (x_{max} \times (1 + \beta), y_{max} \times (1 + \beta))$$
$$D' = (x_{min} \times (1 - \beta), y_{max} \times (1 + \beta))$$

Yellow area which surrounds the vessel candidate region in Fig. 4 presents the initial rectangle region. This region constructed in the first image is used in the second image with the same region. Rectangle region is scanned to the second image inside the rectangle region. If there is a candidate region of blood vessel in this region, its region is processed as a candidate region. If there are no candidates in yellow region, the region is expanded as shown in the blue region in Fig. 4 and candidate region is searched.

Figure 5 shows an example of the determination of candidate region of blood vessel in the second input image. Region surrounded by red line in the Fig. 5 represents a rectangle region that was constituted in the first image. Two vessels candidate region exists in a rectangular area in the second image. In this example, the area of number 1 is 9, while the area of number 2 is 11. The larger region of number 2 is determined as a final candidate region of blood vessel in the second image. In Fig. 5, as a candidate region of blood vessel is observed in the initial rectangle region and the expansion of the rectangle region is not applied.

1.2 The Narrowing of the Blood Vessels Candidate Region

In the second step of the blood vessel detection, the candidate region of blood vessel is detected by color information and labeling, and refinement of candidate region is applied from two input images. Here, the second step of the processing is shown in Fig. 6.

Fig. 5 Example of the determination of candidate region of blood vessel using the rectangle region

0	0	0	0	0	0	0	0
0	0	0	0	0	0	2	2
0	0	0	0	0	2	2	2
0	1	0	0	0	0	2	2
0	1	1	0	0	0	2	0
1	1	1	0	0	2	2	0
0	1	1	0	0	0	2	0
0	0	1	0	0	0	0	0

Fig. 6 Second stage of blood vessel detection

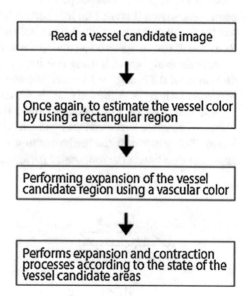

Second stage of vessel detection

Read a vessel candidate image

⬇

Once again, to estimate the vessel color by using a rectangular region

⬇

Performing expansion of the vessel candidate region using a vascular color

⬇

Performs expansion and contraction processes according to the state of the vessel candidate areas

In the second step of detection of blood vessel, color information is obtained from two candidate images of blood vessel detected in the first step. Color of blood vessel is updated by obtaining the color information again although it is obtained once. Candidate of blood vessel is expanded using updated color of blood vessel. Expansion and shrinking processes are applied for the refinement of candidate region of blood vessel after expansion processing. After two candidate regions of blood vessels are obtained, amount of movement of endoscope is estimated by detecting corresponding points by SIFT. Finally the absolute size and shape of polyp are recovered for Lambertian image which is converted from the original image using the previous approach [4, 9] .

To terminate expansion and shrinking process, parameter γ is used to expansion and shrinking as known, here γ is not constant but is changed according to the count of expansion or shrinking processes. The condition to terminate the process becomes stable with this dynamic parameter γ. The initial value of the parameter γ is the same as previous research [10], but value changes according to the number of expansion and shrinking.

Corresponding points are extracted by SIFT [7] using two vessel candidate images detected by expansion and shrinking processing.

1.3 Observation System

Observation system of endoscope image is shown in Fig. 7 under the condition of point light source illumination and perspective projection. Note that it is assumed that the target object is continuous surface with Lambertian reflectance and original RGB endoscope image is converted to the uniform Lambertian reflectance.

Here the situation with using two images as shown in Fig. 7 is considered. The movement of ΔZ is assumed from the correspondence of the observed medical suture between two images. Although triangle which forms the image coordinates x, y and focal length f of the lens becomes similar as the triangle of world coordinates X, Y and Z, absolute shape cannot be obtained without any calibration object in the image. This paper uses the medical suture with known width and proposes a new approach to estimate the reflectance parameter C and the absolute shape of polyp.

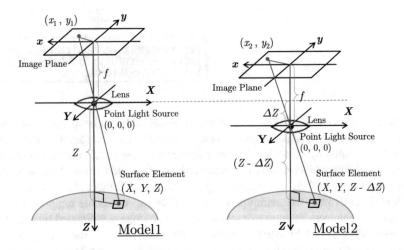

Fig. 7 Observation system

Fig. 8 Observation model of blood vessel

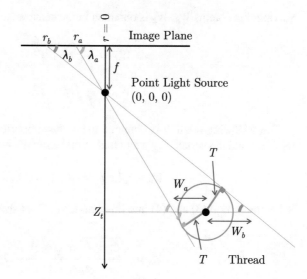

Under the condition that the medical suture is used as the reference object of scaling. Here, a cylindrical model of the medical suture is used for its cross section as shown in Fig. 7.

1.4   Estimation of Movement ΔZ of Camera Along Depth Direction

The amount of movement of endoscope along the depth direction is assumed to be ΔZ. Here, the size of blood vessel is treated as a known constant.

Depth Z_t is obtained for any point of the blood vessel based on the radius T of width. As there is a similarity between a triangle of image coordinate (x, y, f) and a triangle of world coordinate (X, Y, Z), W_a and W_b are first obtained, where W_a and W_b are necessary to calculate Z_t. λ_a is represented from a triangle of image coordinate and a triangle of radius of medical suture and tangent of circumference as follows (Fig. 8).

$$sin\lambda_a = \frac{f}{\sqrt{r_a^2 + f^2}} \qquad (1)$$

$$sin\lambda_a = \frac{T}{W_a} \qquad (2)$$

Solving this obtains W_a. W_b is obtained in the same way.

$$W_a = \frac{T}{f}\sqrt{r_a^2 + f^2} \tag{3}$$

$$W_b = \frac{T}{f}\sqrt{r_b^2 + f^2} \tag{4}$$

The following relation is obtained using the similarity of a triangle of the base $(r_b - r_a)$ and the vertex origin and that of the base $(W_a + W_b)$ and the vertex origin.

$$|r_b - r_a| : f = |W_a + W_b| : Z_t$$

Substituting Eqs. (3) and (4) into this can derive Z_t as follows.

$$Z_t = \frac{\sqrt{r_a^2 + f^2} + \sqrt{r_b^2 + f^2}}{|r_b - r_a|} T \tag{5}$$

Equation (5) represents the depth estimation from one image and this equation can be applied to the depth Z_{t1} of any point of the medical suture in image 1 and the corresponding depth Z_{t2} in image 2. The movement of the camera should be $\Delta Z = Z_{t1} - Z_{t2}$. However this depends on the width of medical suture on the image plane and accuracy of corresponding point between two images.

The approach tries to obtain multiple candidate values of ΔZ using multiple corresponding points between images and estimates the best candidate value of camera movement parameter ΔZ along the depth direction.

Let C_1 be a estimated reflectance parameter for image 1, and let C_2 be that for image2. The best parameter ΔZ is selected from the criteria which minimizes C_r where C_r is calculated from Eq. (6) with the difference between C_1 and C_2 which is estimated from each image 1 and 2, respectively.

$$C_r = \frac{|C_1/C_2| + |C_2/C_1|}{2} - 1 \tag{6}$$

1.5 Determining Initial Value C_{init}

Local brightest point is used to recover the 3-D shape of polyp as an initial point. This is because initial point has constraints where the surface normal vector is the same direction as the light source vector under the Lambertian reflectance. Here the approach first converts into Lambertian image to recovers the 3-D shape of polyp. At the local brightest point under Lambertian reflectance, the relation of $\mathbf{n}$ and $\mathbf{s_1}$ is $\mathbf{n} = \mathbf{s_1}$ and $(\mathbf{s_1}, \mathbf{n}) = \mathbf{1}$.

Let the image coordinate of the initial point in image 1 be (x_1, y_1), let the corresponding coordinate in image 2 be (x_2, y_2), and let the image intensity at (x_1, y_1) be E_1, then the initial candidate of reflectance parameter C_{init} can be given by

$$C_{init} = E_1(X^2 + Y^2 + Z^2)$$
$$= \frac{E_1 Z^2}{f^2} \left(x_1^2 + y_1^2 + f^2\right) \tag{7}$$

C_{init} is derived by substituting Z into Eq. (7) after deriving Z geometrically using the corresponding points between two images.

$$C_{init} = \frac{E_1(\Delta Z)^2}{\{(1 - k^{-\frac{1}{2}})f\}^2} \left(x_1^2 + y_1^2 + f^2\right) \tag{8}$$
$$k = \sqrt{(x_1^2 + y_1^2)(x_2^2 + y_2^2)^{-1}}$$

C_{init} can be uniquely determined by Eq. (8) and it makes possible to obtain the actual scale of object to be recovered.

C can be uniquely determined and the approach makes it possible to obtain the actual scale of object to be recovered.

2 Experiments

Real experiments are done to confirm the effectiveness of detection of blood vessel. Two kinds of real images including those difference images, that is, a total of four images are used to detection of blood vessel. Image size is 1000×870 pixels, size of blood vessel is 0.1 mm, focal length of lens is 5 mm, parameter α is set to be 5%, β is set to be 10%, and initial value of γ is set to be 10%, respectively.

2.1 Blood Vessel Candidate Detection in the Real Image

Input images used in the experiment are shown in Figs. 9, 10, 11 and 12, respectively. Figures 10 and 12 are images after 5 frames of Figs. 9 and 11, respectively.

Next, candidate region of blood vessel was detected by color information and labeling processing from each input image. All candidate regions of blood vessel by labeling in each input image are shown in Figs. 13, 14, 15 and 16, respectively. Table 1 shows the number of candidate region of blood vessel, the maximum area and region number in each image. Results of refinement of blood vessel using expansion

Fig. 9 Input image 1
(pattern 1)

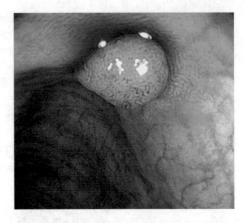

Fig. 10 Input image 2
(pattern 1)

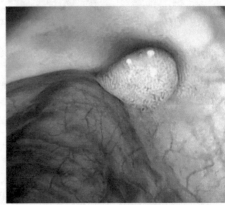

Fig. 11 Input image 1
(pattern 2)

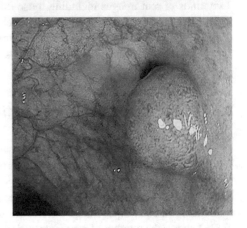

Fig. 12 Input image 2
(pattern 2)

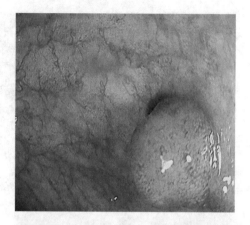

Fig. 13 Candidate of blood
vessel in image 1 (pattern 1)

Fig. 14 Candidate of blood
vessel in image 2 (pattern 1)

Fig. 15 Candidate of blood
vessel in image 1 (pattern 2)

Fig. 16 Candidate of blood
vessel in image 2 (pattern 2)

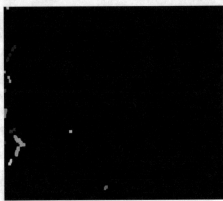

Table 1 Labeling result

Name	Region number	Maximum area
Fig. 9	17	1611 (No.9)
Fig. 10	42	1560 (No.16)
Fig. 11	24	1493 (No.14)
Fig. 12	11	620 (No.8)

and shrinking processing to the detected candidate region are shown in Figs. 17, 18, 19 and 20. Using these results of blood vessel images, amount of movement of endoscope ΔZ was obtained as 3.8342 mm for pattern 1(Figs. 17 and 18), while 1.1764 mm for pattern 2(Figs. 19 and 20), respectively.

Fig. 17 Candidate of blood vessel in image 1 (pattern 1)

Fig. 18 Candidate of blood vessel in image 2 (pattern 1)

Fig. 19 Candidate of blood vessel in image 1 (pattern 2)

Fig. 20 Candidate of blood
vessel in image 2 (pattern 2)

2.2 Shape Recovery of Polyp from Detected Blood Vessel

Experiment is done to estimate the amount of movement ΔZ of endoscope after
obtaining corresponding points by SIFT using two candidate images of blood vessel.

Uniform Lambertian images are generated from the original images with removing specular reflectance are shown in Figs. 21 and 22. Figures 23, 24, 25 and 26 show
the results of recovering 3D shape of polyp based on the approach [12], respectively.
Figures 23 and 24 correspond to the endoscope images of pattern 1 and Figs. 25 and
26 correspond to the endoscope images of pattern 2. Scale unit of vertical axis, horizontal axis and height of these figures is [mm]. In the result of pattern 1, the vertical
size obtained is around 5 mm and horizontal size obtained is around 6 mm, while in
the result of pattern 2, the vertical size obtained is around 7.5 mm and horizontal size
obtained is around 5 mm. Acceptable absolute size for polyp was obtained.

Fig. 21 Generated
Lambertian image (pattern 1)

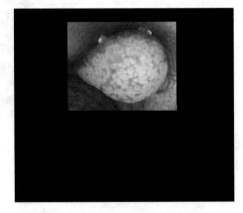

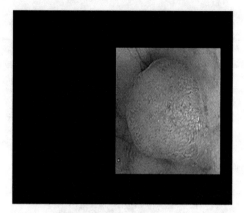

Fig. 22 Generated Lambertian image (pattern 2)

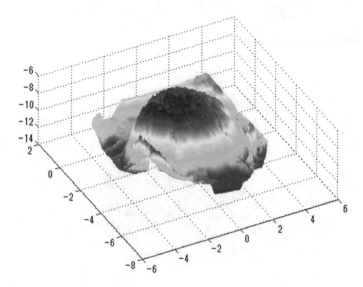

Fig. 23 Recovered shape of pattern 1

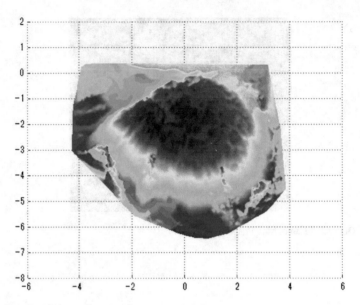

Fig. 24 Recovered shape of pattern 1

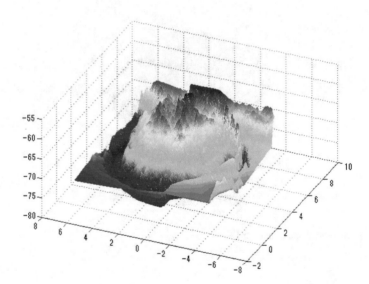

Fig. 25 Recovered shape of pattern 2

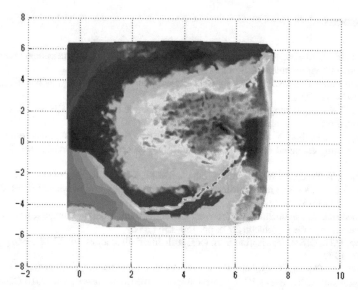

Fig. 26 Recovered shape of pattern 2

3 Conclusion

This paper proposed a new detection approach of blood vessel in endoscope image instead of GrowCut approach. Proposed approach estimates the blood vessel in images using color information and labeling processing. Final candidate of blood vessel is detected by refinement processing with expansion and shrinking. This approach could solve the problem to specify the candidate region of blood vessel with user's operation.

Computer experiments show the usefulness of the proposed approach. The further subjects include the following. This approach detects the blood vessel by color information but sometimes detects the region except blood vessel in some endoscope image. Endoscope image suddenly changes its condition of color information and this is still difficult with the imaging condition. It is also planned to recover the shape of polyp with a variety of endoscope images.

Acknowledgements Iwahori's research is supported by Japan Society for the Promotion of Science(JSPS) Grant-in-Aid Scientific Research(C)(#17K00252) and Chubu University Grant.

References

1. Nakatani, H., Abe, K., Miyakawa, A., Terakawa, S.: Three-dimensional measuremen endoscope system with virtual rulers. J. Biomed. Opt. **12**(5), 051803 (2007)
2. Thormaehlen, T., Broszio, H., Meier, P.N.: Three-dimensional Endoscopy. In: Falk Symposium, pp. 199–212 (2001)
3. Horn, B.K.P.: Obtaining shape from shading information. In: Winston, P.H (ed.) The Psychology of Computer Vision. McGraw-Hill, pp. 115–155 (1975)
4. Neog, D.R., Iwahori, Y., Bhuyan, M.K., Woodham, R.J., Kasugai, K.: Shape from an endoscope image using extended fast marching method. In: Proceedings of IICAI-11, pp. 1006–1015 (2011)
5. Iwahori, Y., Tsuda, S., Woodham, R.J., Bhuyan, M.K., Kasugai, K.: Improvement of recovering shape from endoscope images using RBF neural network. In: Proceedings of ICPRAM 2015, pp. 62–70 (2015)
6. Vezhnevets, V., Konouchine, V.: Grow-cut interactive multi-label N-D image segmentation. In: Proceedings of the Graphicon 2005, pp. 150–156 (2005)
7. Lowe, D.G.: Object recognition from local scale invariant features. In: ICCV 1999, pp. 1150–1157 (1999)
8. Imura, M., Oshiro, O., Chihara, K.: A consideration for extracting continuous component of image using GPU. IPSJ SIG Technical Report 2010 (in Japanese), vol. 2010-CG-138, 11, (2010)
9. Shimasaki, Y., Iwahori, Y., Neog, D.R., Woodham, R.J., Bhuyan, M.K.: Generating lambertian image with uniform reflectance for endoscope image. In: IWAIT2013, pp. 1–6 (2013)
10. Suda, T., Iwahori, Y., Fuhanashi, K., Kasugai, K.: 3D shape recovery of polyp using blood vessel in endoscope images. Meeting on Image Recognition and Understanding 2015 (in Japanese), SS2-16, pp. 1–2 (2015)
11. Iwahori, Y., Yamaguchi, D., Nakamura, T., Kijsirikul, B., Bhuyan, M.K., Kasugai, K.: Estimating reflectance parameter of polyp using medical suture information in endoscope image. In: Proceedings of ICPRAM 2016, pp. 503–509 (2016)
12. Tatematsu, K., Iwahori, Y., Nakamura, T., Fukui, S., Woodham, R.J., Kasugai, K.: Shape from endoscope image based on photometric and geometric constrains. Proced. Comput. Sci. **22**, 1285–1293 (2013)

Design of Agent Development Framework for RoboCupRescue Simulation

Shunki Takami, Kazuo Takayanagi, Shivashish Jaishy, Nobuhiro Ito and Kazunori Iwata

Abstract The RoboCup Rescue Simulation project is one of the responses to recent large-scale natural disasters. In particular, the project provides a platform for studying disaster-relief agents and simulations. The aim of the project is to contribute to soci by making our research findings available. Some of the agents contain excellent algorithms, and so it should be possible to share them among developers. However, this is hindered by the fact that the program structure of the agents is different for each team. Therefore, in this paper, we design and implement an agent-development framework that unifies the structure within the project to facilitate such technical exchange.

Keywords Development platform · Robocuprescue · Rescue simulation · Multi-agent system

S. Takami (✉) · K. Takayanagi · S. Jaishy
Graduate School of Business Administration and Computer Science,
Aichi Institute of Technology, 1-38-1 Higashiyama-dori, Chikusa-ku,
Nagoya, Aichi 464-0807, Japan
e-mail: takamin@maslab.aitech.ac.jp

K. Takayanagi
e-mail: uranos@maslab.aitech.ac.jp

S. Jaishy
e-mail: shivashish@maslab.aitech.ac.jp

N. Ito
Department of Information Science, Aichi Institute of Technology,
1247 Yachigusa, Yakusa-cho, Toyota, Aichi 470-0392, Japan
e-mail: n-ito@aitech.ac.jp

K. Iwata
Department of Business Administration, Aichi University,
4-60-6, Hiraike-cho, Nakamura-ku, Nagoya, Aichi 453-8777, Japan
e-mail: kazunori@vega.aichi-u.ac.jp

© Springer International Publishing AG 2018
R. Lee (ed.), *Computational Science/Intelligence and Applied Informatics*,
Studies in Computational Intelligence 726, DOI 10.1007/978-3-319-63618-4_14

185

1   Introduction

In 2001, the international RoboCup community started the RoboCup Rescue Simulation (RRS) project to confront large-scale natural disasters [1, 7]. In particular, the annual RSS Agent Competition is a platform for studying disaster-relief agents and simulations. Our aim is to contribute to the society by submitting results for this project [12].

However, in order to solve the disaster-relief problems targeted by the RRS, it is necessary to implement a combination of multiple algorithms, such as those for path planning, information sharing, and resource allocation. The Agent Competition is held for the purpose of exchanging technical information and sharing agent programs. Some of the agents contain excellent algorithms, and so it should be possible to share them among developers [3]. However, because program codes differ from one team to the next, it is difficult to re-use them, which hampers technical exchange.

Thus, in this paper, we design and implement an agent-development framework that unifies the structure within the RRS project, thereby allowing program codes to be re-used. This could also act as a platform for researching different disaster-relief algorithms. In the evaluation, we confirm that codes can be re-used.

2   Research and Development in RRS

2.1   Overview of RRS

The RRS is a research platform that simulates disaster-affected areas and disaster-relief activities on a computer [11]. It can handle disaster-relief activities over roughly five hours from the occurrence of a disaster.

Figure 1 shows the activities of agents in the RRS. In the disaster-relief activities, we control six types of agents, namely AmbulanceTeam, FireBrigade, PoliceForce, and the headquarters of each of these units. In addition, there are agents to simulate disaster situations, namely Civilian agents.

- **AmbulanceTeam and AmbulanceCentre**
  These agents rescue other agents that cannot move by themselves.
- **FireBrigade and FireStation**
  These agents extinguish fires in buildings.
- **PoliceForce and PoliceOffice**
  These agents clear road blockages.
- **Civilian**
  In the competition, these agents move automatically to evacuation centers.

The RRS simulator consists of a kernel that manages the progress of the simulation. In addition, there are sub-simulator components that simulate disaster situations and agent programs. The simulation proceeds at the rate of one step per minute.

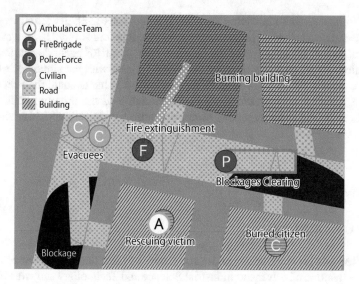

Fig. 1 Overview of RRS

The time allowed for calculating one step of an agent is limited to one second. Before the start of the simulation, as a pre-calculation stage, it is possible to calculate initial data using only topographical information [5].

The RRS can be used to research the application of artificial intelligence and information science to natural-disaster rescue problems. Researchers continue to investigate algorithms for route searching, information sharing, and task allocation in a disaster environment. In the RRS project, five tasks are advocated in particular, namely Group Formation, Path Planning, Search, Multi-task Allocation, and Communication. Every year, competitions using agent programs are held for the purpose of technical exchange [10].

2.2   Agent Development in RRS

The disaster-relief problems handled by the RRS are complex compound problems in which damage situations such as fire, building collapse, and the availability or otherwise of wireless communication change from moment to moment in the various afflicted areas. These changes are addressed by the strategies of teams of disaster-relief robots that differ according to the affected area. To construct a disaster-relief strategy, it is necessary to prepare all the algorithms for tasks such as route searching, information sharing, and resource allocation in the disaster environment. Moreover, in activities such as blockages being cleared by the PoliceForce, it is necessary to use angles and coordinates to specify the direction for activity and the positions of the agents.

To promote research involving the RRS, it is necessary to clarify the structure of a complicated disaster-relief problem and subdivide it before solving it. However, it is problematic that the structure of the program codes for RRS agent development are not unified. At present, if a program can communicate with the simulator's kernel, it can be created freely. However, for researcher to be able to share their respective research fields, it is desirable to unify the structure of the program code and to program each field as a component. Moreover, those components should be re-usable as modules.

2.3 Related Work

2.3.1 OpenRTM-Aist

OpenRTM-aist [2] is a framework for robot development that was developed by the National Institute of Advanced Industrial Science and Technology and whose implementation is based on the RT-Middleware standard [4]. Figure 2 shows a conceptual image of the behavior of RT-Middleware. This common platform standard divides robot elements such as actuator control, sensor input, and algorithms necessary for behavior control into single components that are known as RT-Components (RTC). RT-Middleware then constructs a robot by combining all such RTC. This makes it possible to subdivide the elements that are necessary for controlling the robot. Because each component can be exchanged as a module and existing modules can be included, it is possible to reduce the burden on developers when developing and improving robots.

Because RT-Middleware is applied mainly to real robots, it is suitable for developing robots that are controlled in real time. However, it is difficult to utilize existing code and knowledge when adopting RT-Middleware because RRS agents are programed mainly with a sequential structure.

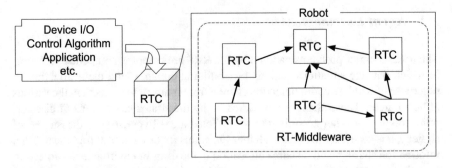

Fig. 2 RT-Middleware behavior

3 Design and Implementation

3.1 Research Objective

In this paper, which is based on the current state of RRS agent development, we implement an agent-development framework by introducing a modular structure to clarify and solve the complicated problems associated with disaster relief. In this way, we make it possible for many researchers to cooperate and solve such problems.

3.2 Design of Agent-Development Framework

For many researchers to clarify and solve complicated problems, we modularize part of the program code. To make it easier to reuse the program code and to reduce the burden on researchers, an agent-development framework is desired. A framework-design method based on OpenRTM-aist (mentioned in Sect. 2.3) would also be conceivable. However, RRS agents are programmed in a sequential structure, so we propose and design a unique Agent Development Framework (ADF) that makes it easier to utilize existing program code and knowledge. This framework facilitates a common architecture for agents, modularization of programs, aggregation of information acquisition interfaces necessary for agent decision-making, and a unified inter-agent communication protocol.

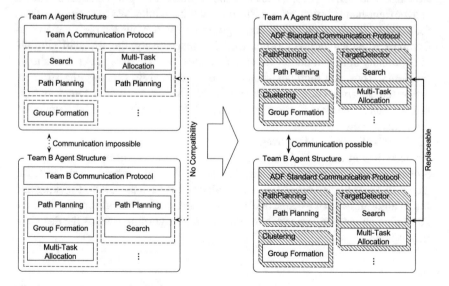

Fig. 3 Before (*left*) and after (*right*) introducing the common architecture

3.2.1 Introduction of a Common Agent Architecture

By defining the overall behavior of an agent as a common architecture, we reduce the differences in combinations of components by each developer and ensure re-usability. This allows developers to implement modules based on this common architecture when developing agent programs.

Figure 3 shows the situations before and after introducing this common architecture. The left-hand portion of the figure shows the existing agent structure, whereas the right-hand portion introduces the common architecture. The portability of the existing program is low because each researcher builds an agent program independently according to individual research agendas. We have commonized the agent-program structure, which is the shaded part of the figure, so that the program code can be re-used easily. Also, this unifies the inter-agent communication protocol and enables communication with agents developed by others.

3.2.2 Modularization of Program Code

As with the RT-Middleware discussed in Sect. 2.3.1, component modularization is introduced to reduce the burden on researchers and to make it possible to reuse the program code used in agent development.

- **Algorithm modularization**
  At the present stage of modularization, we divide as much as possible based on the five tasks presented in the RRS project. Figure 4 shows the relationship between RSS tasks and ADF modules. The left-hand portion of the figure contains five tasks, and the right-hand portion contains the framework modules. We classify algorithms for solving complex problems and algorithms for solving simple problems as Complex Modules and Algorithm Modules, respectively, thereby clarifying the directionality of each module. We view Complex Modules as aggregates of simple problems. Thus, the modules should be programmed by dividing the structure inside the program code as much as possible. Moreover, if it is found that the program able to be divided into modules in the future, we aim to clarify complicated problems as new modules.
- **Modularization of control program**
  As described in Sect. 2.2, it is necessary to control an agent by specifying such properties as its coordinates and angles. Low-level agent control is modularized as a control program. By separating macro algorithms such as decision-making and micro algorithms such as control using the coordinates and angles, we reduce the burden on researchers who wish to study a single algorithm such as the decision-making one.

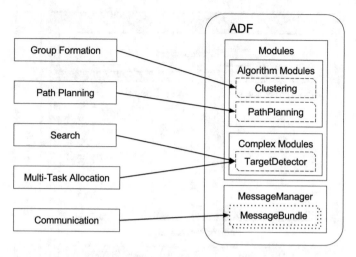

Fig. 4 Relationships between RRS tasks (*left*) and ADF modules (*right*)

3.2.3   Other Approaches

The introduction of a common architecture makes it possible to aggregate information-acquisition interfaces, manage parameters collectively, and unify the inter-agent communication protocol.

- **Aggregation of information-acquisition interfaces**
  The framework aggregates the interfaces that acquire information necessary for agent decision-making provided by the kernel. This clarifies the data that are acquired.
- **Collective management of parameters**
  The framework collectively manages the specifications of the modules to be used, the eigenvalues in the algorithm that is to be changed at the time of the experiment, and the pre-computed data. This makes it easier to manage all the various parameters.
- **Unifying inter-agent communication protocols**
  In the RRS, the inter-agent communication protocols are currently not unified. This strengthens the dependency between components but makes it difficult to modularize the algorithm. Therefore, with reference to the RRS inter-agent communication protocols proposed by Ota et al. [9] and Obashi et al. [8], we define a common inter-agent communication protocol. We define messages communicated under this protocol as members of either an information-sharing family or a command family. In the information-sharing family, information about agents, roads, and buildings is shared. In the command family, commands for relief, fire extinguishing, blockage clearing, and searching are commanded.

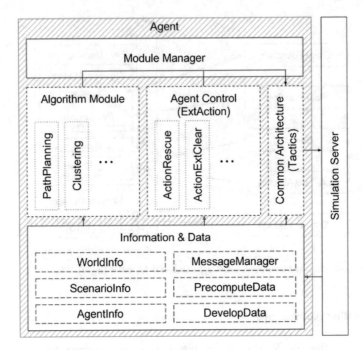

Fig. 5 Structure of framework

3.3 Implementation of Agent-Development Framework

Figure 5 shows the structure of the agent-development framework that is implemented. The internal aspects of the agent, as represented by the shaded part in the figure, are implementations of the framework. The parts indicated by dotted lines are modules, which are the objects to be programmed at the time of agent development. As shown in Fig. 6, the agents' thinking process has a state in which all modules are basically implemented by programming the processing for each state. In this section, we describe each implementation of the framework based on the design discussed in Sect. 3.2.

3.3.1 Common Agent Architecture

This component defines how agents invoke algorithm modules and control-program modules to determine their behavior. This component is referred to as Tactics. Each agent determines its behavior by invoking elements such as task assignment, a route-search algorithm, and a control-program module within Tactics. In addition, each module is invoked by its alias name and can select the module to be attached by the agent-configuration file.

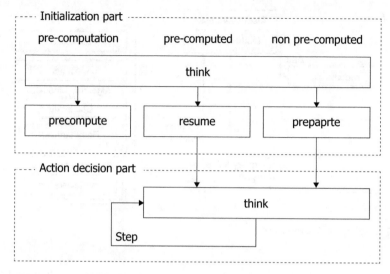

Fig. 6 Stage of the agents' thinking

Each action is expressed by the following class instances, and they are determined by returning it to the code that called the Think() method.

- **ActionRest**
  Stay in place
- **ActionMove**
  Move along the path (Arg.: Path, [X], [Y])
- **ActionRescue (for AmbulanceTeam)**
  Rescue the victim (Arg.: TargetVictim)
- **ActionLoad (for AmbulanceTeam)**
  Accommodate the victim (Arg.: TargetVictim)
- **ActionUnload (for AmbulanceTeam)**
  Set down the victim
- **ActionExtinguish (for FireBrigade)**
  Extinguish fire (Arg.: TargetBuilding)
- **ActionRefill (for FireBrigade)**
  Refill tank with water
- **ActionClear (for PoliceForce)**
  Clear blockage on road (Arg.: X, Y)

3.3.2   Modules

As shown in Fig. 4 (Sect. 3.2.2), each task is modularized as Clustering, PathPlanning, or TargetDetector. Also, the control program is modularized as ExtAction.

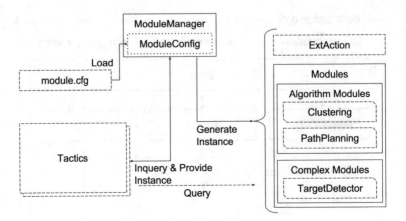

Fig. 7 Loading module instances

An instance of the module is created and managed in a component called the ModuleManager. This further enables switching the module to be used by loading the module-configuration file (*module.cfg*) at agent startup.

3.3.3 Collective Management of Parameters

Figure 7 shows the flow of loading module instances. The settings of the module to be loaded are obtained from the aforementioned file *module.cfg*. Parameters in algorithms can be obtained from the DevelopData component, and each value can be input by JSON [6] formatted text as an argument at agent startup. In addition, the pre-computed data can be stored in the PrecomputeData class.

3.3.4 Inter-agent Communication Protocols

We have defined messages as members of either an information-sharing family or a command family. The following messages are implemented in an information-sharing family:

- **MessageRoad**
  Contains the state of a road and includes information about any blockages.
- **MessageBuilding**
  Contains the state of a building.
- **MessageCivilian**
  Contains the state of a civilian.
- **MessageAmbulanceTeam**
  Contains the state of the AmbulanceTeam and includes the actions of the agents.

- **MessageFireBrigade**
  Contains the state of the FireBrigade and includes the actions of the agents.
- **MessagePoliceForce**
  Contains the state of the PoliceForce and includes the actions of the agents.

The following message is implemented in a command family. The distinction between the commands and the request is judged by whether there is a broadcast designation.

- **CommandAmbulance**
  Sends commands and requests to the AmbulanceTeam.
- **CommandFire**
  Sends commands and requests to the FireBrigade.
- **CommandPolice**
  Sends commands and requests to the PoliceForce.
- **CommandScout**
  Sends scout commands and requests.
- **MessageReport**
  Reports the results of the commanded actions.

For communication, as shown in Fig. 5, we use MessageManager in "Information and Data." We register the message to be sent to MessageManager for transmission. The received messages retrieve from MessageManager. Specifying the class of a message allows us to retrieve only specific messages if we so wish.

4 Example of Implementation of Agent Using ADF

In this section, we describe an example of an agent that implements a unique algorithm for the PoliceForce route-searching algorithm. Every other module can be implemented by following an equivalent procedure.

4.1 TacticsPoliceForce

The common architecture TacticsPoliceForce implements the invoking structure of the module shown in Fig. 8. The shaded area is the part that processes the command from headquarters. The agents that used this TacticsPoliceForce always act in a decentralized way because those shaded modules give a null return as a result in this implementation. The other modules handle autonomous behavior. This TacticsPoliceForce is currently a common architecture for PoliceForce in the whole project.

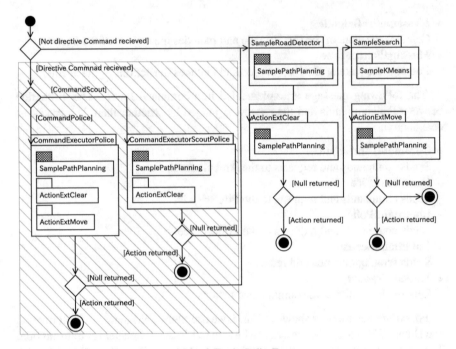

Fig. 8 Module-invoking structure of SampleTacticsPoliceForce

4.2 PathPlanning

Developers implement a unique path-planning algorithm, details of which are not described here because it is different from the essence of this paper.

As shown in Fig. 5, the inputs of the module are the starting location, ending location, and the components shown in "Information and Data". Implementation programs calculate routes based on the input into the calc() method. We implement the getResult() method so that the route path can be acquired.

4.3 Loading Modules

Module incorporation changes the specification of PathPlannig in the file *module.cfg*. The specification of the module that undergoes changes are as follows:

- CommandExecutorPolice.PathPlanning
- CommandExecutorScoutPolice.PathPlanning
- ActionExtClear.PathPlanning
- ActionExtMove.PathPlanning
- (SampleRoadDetector.PathPlanning)

5 Evaluation

5.1 Purpose of Evaluation

With regard to the agent-development framework proposed in Sect. 3, we developed and tested the agent program at an actual workshop to evaluate the re-usability of the code. We collaborated with multiple developers with different programming philosophies to assess whether the agents could be developed in the proposed environment.

5.2 Experiments

At the workshop of the RoboCup Simulation league held at Fukuoka University on October 22–23, 2016, roughly 30 people were divided into six teams to develop agents. We conducted an experiment to collaborate and work in combination with AmbulanceTeam, FireBrigade, and PoliceForce agents that were developed by each team. In doing so, we confirmed that it was possible in this experiment to combine agents developed by different developers. At the workshop, we experimented with only six combinations because of time restrictions. However, we have since experimented with a total of 216 combinations.

Table 1 Top eight results of combined experiments (* indicates reference values)

Rank	AT	FB	PF	Score
1	A	C	A	143.4091
2	A	B	A	141.4178
3	E	B	A	140.9280
4	A	B	F	136.5078
5	E	B	B	134.8257
6	B	B	D	134.3782
7	C	B	D	134.3404
8	B	B	E	133.8126
(*)	Sample	Sample	Sample	114.2580
	A	A	A	114.2574
	B	B	B	126.5861
	C	C	C	124.3933
	D	D	D	114.2580
	E	E	E	119.9167
	F	F	F	120.8222

5.3 Results and Discussion

Table 1 gives the top eight results of the combined experiments of each team (A–F), those of a sample team, and the results for each team as a reference. The score of each team is no better than the combination of the top eight because the time for development was short. However, because the scores are high when these agents are combined, it can be seen that agents developed by different developers are cooperative. We confirmed that the agents developed on the proposed framework could work in conjunction with each other. The reason why developers could develop agents in a limited time was because of cooperation within each team. This confirms that the codes are re-usable.

6 Conclusion

In this paper, we designed and implemented an agent-development framework that unifies the structure within the project to foster technical exchange. In the evaluation, we confirmed that the codes could be re-used. Also, we confirmed that agents could be developed by multiple developers in collaboration. As mentioned in Sect. 3.2.2, the algorithms for solving complex problems (known as the Complex Modules) are aggregated from simpler problems. Thus, in the future, if it is found empirically that a particular program can be divided, we aim to clarify complicated problems as new modules. This will lead to better resolution of disaster-relief problems. Eventually, we aim to return the research results of the RRS project to society by clarifying disaster-relief problems and proposing individual algorithms that are applicable to disaster relief.

Acknowledgements This work was supported by JSPS KAKENHI Grant Number JP16K00310 and JP17K00317.

References

1. Robocuprescue simulation. http://roborescue.sourceforge.net/
2. N.I. of Advanced Industrial Science Technology: Openrtm-aist. http://www.openrtm.org/
3. Akin, H.L., Ito, N., Jacoff, A., Kleiner, A., Pellenz, J., Visser, A.: Robocup rescue robot and simulation leagues. AI Mag. **34**(1), 78–86 (2013). http://dblp.uni-trier.de/db/journals/aim/aim34.html#AkinIJKPV13
4. Ando, N., Suehiro, T., Kitagaki, K., Kotoku, T., Yoon, W.: Rt-middleware: distributed component middleware for rt (robot technology). In: 2005 IEEE/RSJ International Conference on Intelligent Robots and Systems, pp. 3933–3938 (2005). doi:10.1109/IROS.2005.1545521
5. Faraji, F., Nardin, L.G., Modaresi, A., Helal, D., Ito, N.: Robocup rescue simulation league agent 2016 competition rules and setup. http://roborescue.sourceforge.net/web/2016/downloads/rules2016.pdf

6. The JSON data interchange format. Technical Report Standard ECMA-404 (1st edn.) / October 2013, ECMA (2013). http://www.ecma-international.org/publications/files/ECMA-ST/ECMA-404.pdf
7. Kitano, H., Tadokoro, S.: Robocup rescue: a grand challenge for multiagent and intelligent systems. AI Mag. 22(1), 39 (2001)
8. Obashi, D., Hayashi, T., Iwata, K., Ito, N.: An implementation of communication library among heterogenous agents naito-rescue 2013 (Japan). In: RoboCup 2013 Eindhoven (2013)
9. Ohta, T., Toriumi, F.: Robocuprescue2011 rescue simulation league team description. In: RoboCup 2011 Istanbul (2011)
10. Skinner, C., Ramchurn, S.: The robocup rescue simulation platform. In: Proceedings of the 9th International Conference on Autonomous Agents and Multiagent Systems, AAMAS'10, vol. 1, pp. 1647–1648. International Foundation for Autonomous Agents and Multiagent Systems, Richland, SC (2010). http://dl.acm.org/citation.cfm?id=1838206.1838523
11. Takahashi, T., Takeuchi, I., Koto, T., Tadokoro, S., Noda, I.: Robocup rescue disaster simulator architecture. In: RoboCup 2000: Robot Soccer World Cup IV, pp. 379–384. Springer, London, UK (2001). http://dl.acm.org/citation.cfm?id=646585.698826
12. Visser, A., Ito, N., Kleiner, A.: RoboCup rescue simulation innovation strategy, pp. 661–672. Springer International Publishing, Cham (2015). doi:10.1007/978-3-319-18615-3_54

Mist Computing: Linking Cloudlet to Fogs

Minoru Uehara

Abstract Recently, cloud computing has become popular and edge computing technology is hence developing. Cloudlets and fogs are concepts proposed in edge computing. The edge-most cloudlet is far from the centered cloud, and the range of a fog
is widespread. In this paper, we propose placing a mist between cloudlets and fogs.
The mist is a data center of cloudlets and a fog device. We describe the requirements
and functions of mist computing. In addition, to show the usefulness of mist computing, we implement a Managed Network Block prototype. We demonstrate that it
is easy to develop Internet of Things devices using the mist framework.

1 Introduction

Recently, cloud computing [1, 2] has become popular and Internet of Things (IoT) [3]
is hence developing. In cloud computing, users use data centers through the Internet
and they can process big data using the large resources of a data center. In the IoT,
everything connects to the Internet. Here, the Internet means the cloud. An essential
feature of creating the IoT is to make things smart. The cloud provides smart services
to IoT. Therefore, the IoT is also known as the Cloud of Things (CoT) [4, 5].

Furthermore, an essential feature of a CoT is device management. A cloud provides smart services to devices. For this purpose, it needs to manage devices. In
this paper, a managed device is defined as a device managed by a cloud. Many
IT providers provide cloud services to managed devices. However, most services
manage individual devices and do not support the federation of devices.

IFTTT [6] and Bluemix [7] realize a federation of devices by combining multiple
services. However, to do so, special skills such as programming are required. It is
hard for beginners to use such services.

We are developing a service called Managed Network Blocks (MNBs) that can
combine network devices freely, like LEGO blocks. A managed network service
(MNS) [8, 9] is a service that manages networks. However, MNSs are suited for large-

M. Uehara (✉)
Department of Information Sciences and Arts, Toyo University, Kawagoe, Japan
e-mail: uehara@toyo.jp

© Springer International Publishing AG 2018   201
R. Lee (ed.), *Computational Science/Intelligence and Applied Informatics*,
Studies in Computational Intelligence 726, DOI 10.1007/978-3-319-63618-4_15

scale networks, not small-scale networks. Small and Medium Businesses (SMBs) use small-scale networks. Home networks are also small scale. In small-scale networks, it is difficult to install an MNS because of its high cost. In contrast, an MNB is suited for SMBs and home networks.

A device such as an MNB is placed between a cloudlet and the fog. We call such device a mist. A mist is larger than a cloudlet (droplet) and smaller than a fog. Mist computing bridges cloudlets and fogs. In this paper, we describe the concepts of mist computing, its requirements, and basic design. In addition, we implemented a simple MNB prototype and define the basic framework.

This paper is organized as follows. In Sect. 2, we survey several related works. In Sect. 3, we propose mist computing and describe its concept. In Sect. 3, we describe the MNB as an example of mist computing. In Sect. 5, we describe the mist-based web services of MNB. Finally, we conclude the paper in Sect. 6.

2 Related Works

2.1 Edge Computing

Edge computing is computing that occurs at the edge of the Internet. In this context, the edge is the frontier of applications. The Internet has grown drastically since it was created, and will continue to grow for the foreseeable future. Even if the Internet covers everywhere on the earth, it will grow in density and any time. Currently, everything is becoming connected to the Internet, which forms the IoT. The frontier of the Internet is edge computing. In edge computing, various devices that have never connected to the Internet will do so.

Fog computing and cloudlets are instances of edge computing. Cisco Systems, Inc., coined the term fog computing [10]. A fog is a cloud close to the ground (i.e., users). In fog computing, a fog consists of edge devices and their networks. According to [10], a fog covers not only Local Area Networks (LANs) but also small things networks (STNs) and field area networks (FANs). In an STN, an IP-incompatible protocol such as Bluetooth is used, and in a FAN, an IP-compatible protocol such as Wi-Fi is usually used. In the upper stream of a FAN, the LAN overlaps fog and cloud computing. To avoid conflict, in this paper, we define fog computing and cloud computing as follows.

- Fog computing is FAN/LAN based computing.
- Cloud computing is Wide Area Network (WAN) based computing.

Cloud computing is sometimes called computing by Internet. However, data centers connected to the Internet perform the computing. Therefore, we define fogs and clouds as follows.

- A fog is the data center of fog computing in a FAN/LAN.
- A cloud is the data center of cloud computing in a WAN.

In this sense, a fog is similar to a private cloud.

A cloudlet [11–13] is a data center for small devices. The term cloudlet means a small cloud, and it is used because small devices regard a cloudlet as a cloud. For example, an iPhone is the data center of an Apple Watch. Therefore, the iPhone is the cloudlet of the Apple Watch. Using this taxonomy, a cloudlet is defined as follows.

- A cloudlet is a data center in an STN.

In addition, Wang [14] and Skala et al. [15] proposes dew computing. Dew computing is similar to cloudlet. A cloudlet connects smart objects with IP-incompatible protocol such as ZigBee [16]. There is large gap between cloudlets and fogs. To fill into the gap, we propose mist computing. A mist is a small fog.

- A mist is a data center for cloudlets in a FAN.

We summarize the definitions as follows.

- STN: Sensors → Cloudlet
- FAN: Cloudlets → Mist
- LAN: Mists → Fog
- WAN: Fogs → Cloud

2.2 Cloud of Things

Parwekar stated that the future of the IoT is the CoT [4], and Aazam et al. believe that fog computing and smart gateways are important for the CoT [17]. Most IoT devices cannot directly connect to the cloud; they require an ARM Linux board. However, most IoT devices are smaller than an ARM Linux board. A smart gateway takes data from small IoT devices and relays it to the cloud. The mist proposed in this paper is similar to a smart gateway.

Indeed, most small IoT devices cannot communicate with a cloud by themselves. To communicate with a cloud, an IoT device usually needs to run ARM Linux. However, most IoT devices are smaller than an ARM single-board computer. Sometimes, a smart gateway is based on ARM Linux. A smart gateway receives data from small IoT devices and forwards them to the cloud at the best data rate and time. The mist that we proposed in this paper is also a smart gateway.

In the IoT, various protocols are used. It is difficult for a smart gateway to support all protocols. Hence, more than one smart gateway will be required, and the network will be more complex.

In previous work, we proposed the framework of the CoT and developed a simple CoT device [5]. Figure 1 shows an overview of the CoT system. In this prototype, we employ an Arduino [18] as a sensor driver and a Raspberry Pi [19] as a smart

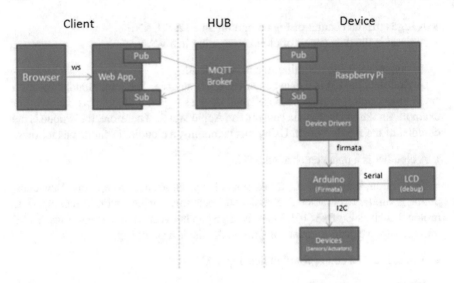

Fig. 1 Overview of a CoT system

gateway. The protocol used between the sensor drivers and smart gateway may be an original protocol. In addition, one issue when setting up a smart gateway in the home is access over a Network Address Translation (NAT). A home network is usually a private network. Therefore, communication is unidirectional. A NAT allows access from inside the network to outside but denies access to the outside from inside the network. However, we can implement bidirectional access over NAT using MQTT [20].

3 Mist Computing

3.1 Concepts of Mist Computing

As mentioned in Sect. 2.1, edge computing can be classified using the distance from the center of the cloud. A cloudlet is the farthest from the cloud. Fog computing covers the region from a FAN to a LAN. However, it is difficult to cover both networks using the same method. Hence, we re-define fog as the data center of a LAN and newly define a mist as the data center of a FAN.

In mist computing, there are three elements: clouds, mists and droplets. A cloud manages more than one mist. A mist manages more than one droplet, a droplet is a device managed by a mist, a droplet is cloudlet of managed devices, and a mist is cloudlet of droplets.

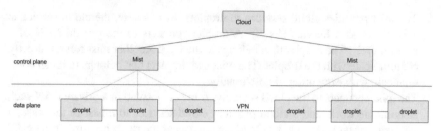

Fig. 2 Hierarchy in mist computing

In addition, mist computing is also a framework for realizing MNBs. Figure 2 shows an overview of mist computing. In Fig. 2, a mist controls several droplets and a cloud controls two mists. The control plane is the layer for control. A mist governs the local control plane and a cloud governs the global control plane. Droplets organize the data plane. The structure of the data plane depends on the application. In this paper, to realize an MNB, the data plane organizes the network. The relationship between the control and data planes is based on OpenFlow [21].

3.2 Cloud

A cloud manages the mists and manages droplets through mists. Users use mists and droplets through a cloud.

3.3 Mist

Mist Requirements

The requirements of a mist are as follows.

1. The mist has a (wired) WAN network interface card (NIC) that connects it to the Internet.
2. The mist has a (wireless) LAN NIC that connects it to droplets. This LAN is a private network and should be wireless.
3. The mist authorizes droplets. It checks whether a droplet that connects to the LAN is registered or not. If a droplet is not registered, its network address must be isolated. For a wireless LAN, a multiple Service Set Identifier (multi-SSID) can organize more than one network. One network is a private LAN to which a registered droplet connects and the other is a public LAN to which an unregistered droplet connects. Both use private network addresses. This allows a droplet authorized in the public LAN to connect to the private LAN.

4. The mist provides cloud services to droplets. In this case, the cloud service is a Software as a Service (SaaS). A droplet connects to the private LAN of a mist securely. The droplet then belongs to the mist. Another mist cannot directly communicate with the droplet. The mist and droplets that belongs to it can communicate with each other bidirectionally.
5. The mist connects to the cloud securely. A mist is located in a private LAN such as home network, and a cloud provides cloud services to mists. In this case, a mist can access a cloud but a cloud cannot access the mists because of the NAT of the private LAN.
6. The mist runs a mist web service, providing mist web services to droplets and the cloud.

Mist Behavior

The behavior of a mist is as follows.

1. The mist activates the WAN. The WAN port is reachable by the Internet, and Dynamic Host Configuration Protocol (DHCP) runs in the WAN.
2. The mist connects to a cloud. When the mist is first booted, it registers itself with the cloud, which activates it. At every boot, it notifies the cloud that it is running. Periodically, it synchronizes its states (e.g., SSID and KEY) with the cloud.
3. The mist activates the LAN. The mist has two LANs. One is a private LAN to which a registered droplet connects. Another is a public LAN to which an unregistered droplet connects. The public LAN cannot communicate with the private LAN. The private LAN connects with the data plane. A mist provides DHCP to the LAN. When a droplet connects to the private LAN, DHCP assigns a private address to the droplet. Note that here, the private network addresses of the control and data planes must be different. It is forbidden to forward data between the LAN and WAN. Therefore, IPv6 is suited for the LAN.
4. The mist activates the mist web service. An HTTP server and MQTT [20] server cooperate to provide the mist web service, which provides an authorized application programming interface (API) for the private LAN and authentication API for the public LAN.

Mist Web Services

A mist web service has the following functions.

1. The web service of a mist registers a droplet. An owner registers a droplet with the cloud. A new droplet connects to the public LAN and registers itself with the mist. The mist asks the cloud whether the droplet is registered or not. If the owners of the mist and droplet are equivalent, the mist registers the droplet. A mist gives the registered droplet access information for the private LAN.
2. The web service of the mist authorizes a droplet. The mist gives a droplet access information for the private LAN. A droplet configures itself using this access information.

3. The web service of a mist gets/sets the state of a droplet. The mist gets the current state of a droplet and sets the new state of a droplet using the droplet web service.
4. The web service of a mist processes the requests of the cloud. This service obtains requests from the cloud. If there is a request, it processes them as follows. If the request is a request for the mist, the mist processes the request by itself. If the request is a request for a droplet, the mist sends it to the droplet and waits for the response from the droplet. The mist receives the response and then sends back it to the cloud.

There are two methods for obtaining requests from the cloud. One is polling, where the mist checks for requests periodically. It is easy to implement polling using cron; however, the polling interval of cron is at least 1 sec. In addition, unnecessary traffic is increased. The other method is notification. In this case, a sender (cloud) notifies a receiver (mist). We can easily implement the notification mechanism using MQTT, which is based on a publish/subscribe model. Once the mist subscribes to the cloud, the cloud can notify the mist whenever it publishes something.

3.4   Droplet

Droplet Requirements

The requirements of a droplet are as follows.

1. The droplet has a (wireless) LAN NIC to communicate with the mist. This LAN is private network and should be wireless.
2. The droplet runs a droplet web service that it provides to the mist.

Droplet Behavior

The behavior of a droplet is as follows.

1. The droplet activates the LAN. The LAN is reachable by a mist, and in the LAN, a mist provides DHCP.
2. The droplet connects to the public LAN. It registers itself with the mist web service through the public LAN. Then, after the mist authorizes it, the mist gives it access information for the private LAN.
3. The droplet connects to the private LAN. It sets up and activates the private LAN connection, then switches from the public LAN to the private LAN.
4. The droplet activates droplet web service using an HTTP server.

Droplet Web Service

A droplet web service has the following functions.

1. The web service of the droplet gets/sets its own state. The droplet web service accepts the requests of the mist web service and gets and sets the state of the droplet. The state of a droplet is represented by a set of properties, which are defined to be a property name and value pair. A property binds a physical resource. If a property value is changed, the physical resource is also changed. Some properties are read only. If anyone tries to change them, an error will occur.
2. The web service of the droplet notifies the mist of its own state change. The web service of the droplet always observes its own states. If needed, it notifies the mist of a state change. A mist can choose a set of properties to observe. In a droplet, the physical resource is periodically observed using cron. If a change is found, the droplet notifies the mist web service. In such a case, cron generates network traffic.

4   Managed Network Blocks

Here, we describe Managed Network Blocks (MNB) as an example of mist computing. MNB is network device that can be combined with other blocks, similar to LEGO blocks.

If the number of IoT devices increases in future, home networks will be more complex than they are today. Currently, a typical home network simply consists of a Wi-Fi access point (AP) connected by Fiber to the Home (FTTH) or Asymmetric Digital Subscriber Line (ADSL). However, such an AP supports at most 30 clients simultaneously. Hence, more than 30 IoT devices cannot connect to a home network. Therefore, even a beginner must organize the home network by combining a wireless network with a wired network.

Furthermore, if IoT devices increase in number, it is difficult to detect failed devices. In network, there are two kinds of failures, node failure and link failure. To detect failures, we determine the network topology and check whether each device is working normally or not. A link failure may divide a network into two subnetworks. It is impossible to check nodes behind a failed link.

To solve these issues, we propose MNB. Currently, we are developing MNB using a Raspberry Pi [19]; however, we have not completed it yet.

There are several kinds of devices in an MNB. Table 1 shows the MNB types. Figure 3 shows an example of a network organized using MNB. The top-rightmost NAT router connects to a WAN. PC2-4 can access PC1 and PC3-4 is connected to each wireless network. The wireless network that PC4 connects with is bridged.

Table 1 Types of MNB

Type	Function
Router	Wired router. DHCP gives WAN address. LAN address of the router itself is a fixed private address given manually. DHCP gives LAN addresses
NAT router	Wired NAT (Network Address Translation) router. DHCP gives WAN address. LAN address of the router itself is a fixed private address given manually. DHCP gives LAN addresses. NAT is enabled
Bridge(L4)	4 port wired bridge
Bridge(L2)	2 port wired bridge
Bridge(LW)	1 port wired-to-wireless bridge
Bridge(WW)	Wireless-to-wireless bridge

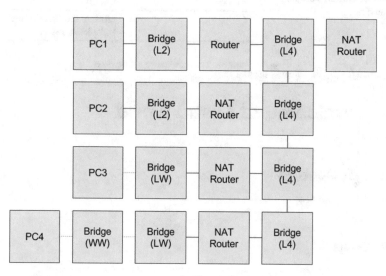

Fig. 3 MNB example

5 A Prototype of Web Services

Here, we describe an implementation of a mist web service and a prototype of MNB. We implemented the mist and droplet web services for MNB using Sinatra [22] as RESTful services on an instance of Vagrant [23]. We show a simple example as follows.

Table 2 shows the list of routes provided by the mist web service, and Table 3 shows the list of routes provided by the droplet web service.

Table 2 Routes in a mist web service

Route	Action
/ Top	
/attach ?url	Register a droplet specified by url
/devices	Return the list of droplets registered to the mist
/devices/:id	Return the states of a droplet specified by id
/devices/:id/detach	Unregister a droplet specified by id

Table 3 Routes in a droplet web service

Route	Action
/	Return the states of the droplet as JSON

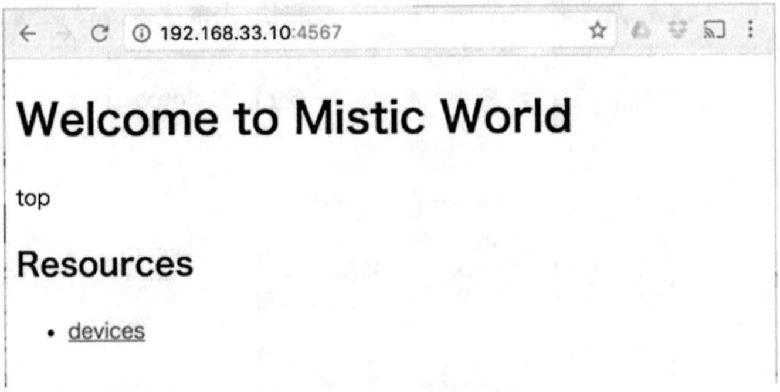

Fig. 4 Top page

Fig. 5 Initial list without droplets

Fig. 6 Droplet attached to the mist

Fig. 7 State of a droplet

First, the mist web service starts. The mist web service uses port 4567, which is the standard WEBrick port used in Sinatra. Figure 4 shows the top page of the mist web service. Indeed, this example is a web application rather than a web service. However, the APIs are used in this application. We can show the list of droplets by following the devices link.

Figure 5 shows the list of droplets. Initially, there are no droplets in the list.

Here, we activate device-0 to start the droplet web service. Device-0 attaches to the mist. As a result, the mist web service registers device-0. Figure 6 shows the list of devices attached to the mist. The list now includes device-0.

Next, we show the states of device-0. The mist gets the state of device-0 by accessing its droplet web service. Figure 7 shows the actual state of device-0.

6 Conclusions

In this paper, we proposed mist computing, which fills the gap between cloudlets and fogs, and we designed a concrete framework for it. A mist is a kind of cloud that is smaller than fog but larger than a cloudlet. Therefore, it is suitable as a layer between a cloudlet and fog. In this paper, to show the usefulness of mist computing, we discussed how to implement an MNB using it. The framework of mist computing includes access over NAT. In addition, an MNB can organize networks freely. Using a mist, we can easily develop the IoT. As a future work, we will complete the MNB implementation.

References

1. Mell, P., Grance, T.: The NIST Definition of Cloud Computing (2011). NIST Special Publication 800-145(draft)
2. Juhnyoung, L.: A view of cloud computing. Int. J. Netw. Distrib. Comput. 1(1), 2–8 (January 2013)
3. Ashton, K.: That 'Internet of Things' thing. RFID J. (2009)
4. Parwekar, P.: From internet of things towards cloud of things. In: 2011 2nd International Conference on Computer and Communication Technology (ICCCT), pp. 329–333, 15–17 Sept 2011. doi:10.1109/ICCCT.2011.6075156
5. Uehara, M.: A case study on developing cloud of things devices. In Proceeding of the 9th International Conference on Complex, Intelligent and Software Intensive Systems (CISIS2015), pp. 44–49, 8–10 July 2015, Regional University of Blumenau (FURB), Blumenau, Brazil
6. IFTTT http://ifttt.com/
7. IBM Bluemix https://www.ibm.com/cloud-computing/bluemix/
8. Hori, K., Yoshihara, K., Horiuchi, H.: Customer equipment configuration manager for managed network service providers. In: 2007 10th IFIP/IEEE International Symposium on Integrated Network Management, Munich, 2007, pp. 516–526. doi:10.1109/INM.2007.374816
9. Rahman, N.A.A., Azmat, F., Yusof, M.I.: Hybrid optimisation for managed network services. In: 2014 IEEE 5th Control and System Graduate Research Colloquium, Shah Alam, 2014, pp. 141–146. doi:10.1109/ICSGRC.2014.6908711
10. Bonomi, F., Milito, R., Zhu, J., Addepalli, S.: Fog computing and its role in the internet of things. In: Proceedings of the 1st edn. of the MCC workshop on Mobile cloud computing (2012)
11. Satyanarayanan, M., Bahl, P., Caceres, R., Davies, N.: The case for VM-Based cloudlets in mobile computing. IEEE Pervasive Comput. 8(4), 14–23 (2009) Oct-Dec
12. Satyanarayanan, M: Mobile computing: the next decade, ACM SIGMOBILE Mobile Comput. Commun. Rev. Arch. 15(2), 2–10 (April 2011)

13. Quwaider, M., Jararweh, Y.: Cloudlet-based for big data collection in body area networks. In: Internet Technology and Secured Transactions (ICITST), 2013 8th International Conference for, pp. 137–141. London (2013)
14. Wang, Y. (2015) Cloud-dew architecture, Int. J. Cloud Computing, Vol. 4, No. 3, pp.199210.
15. Skala, K., Davidovic, D., Afgan, E. Sovic, I. Sojat, Z.: Scalable distributed computing hierarchy: cloud, fog and dew computing, Open J. Cloud Comput. (OJCC), 2(1), 16–24, (2015)
16. Chmaj, G., Selvaraj, H.: Energy-efficient computing solutions for internet of things with zigbee reconfigurable devices. Int. J. Softw. Innovation (IJSI) 4(1)
17. Aazam, M., Eui-Nam Huh: Fog computing and smart gateway based communication for cloud of things. In: 2014 International Conference on Future Internet of Things and Cloud (FiCloud), pp. 464–470, 27–29 Aug 2014, doi:10.1109/FiCloud.2014.83
18. Arduino: arduino—home, https://www.arduino.cc/
19. Raspberry pi foundation: raspberry pi—teach, learn, and make with raspberry pi, https://www.raspberrypi.org/
20. MQTT http://mqtt.org/
21. McKeown, N. et al.: OpenFlow: enabling innovation in campus networks. ACM SIGCOMM Comput. Commun. Rev. 38(2), 69-74 (2008)
22. Sinatra http://www.sinatrarb.com/
23. HashiCorp: vagrant by hashiCorp https://www.vagrantup.com/

Self-recognition and Fault Awareness in OpenFlow Mesh

Suguru Yasui and Minoru Uehara

Abstract Energy conservation is important to protect available resources in the world. Computer technology is the same. In order to realize this idea in computer technology, renewable computing will be important. It is desirable for a computer to adopt technologies that are suitable to their era and environment in order to reduce their energy use. However, many computers used in business are large. It is difficult for them to adopt such technologies dynamically. For this reason, their computers are not capable of reducing their energy usage at this time. Therefore, we propose a computing system that adapts to environments and whose continued development will be made possible by replacing faulty and aging component in the system dynamically. A system whose components can easily be replaced will be achieved using the metabolic architecture. In this paper, we implement the two functions of self-recognition and fault awareness for a calculating unit in the system.

1 Introduction

In recent times, renewal computing is becoming more important. According to [1], renewal computing is divided into two types: rebirth, and immortality. Rebirth corresponds to a like replacement, while immortality corresponds to sustainability computing and normally-off computing [2]. Sustainable computing does not mean only renewable energy based systems, but also suggests environmental adaption and continuous development. As architecture of such a system, we propose metabolic architecture [3, 4]. Metabolic architecture always maintains the new condition of the computer by aggressively exchanging its components. This characteristic is available to large systems, such as online cloud technology. We will implement the ARM Linux Cloud as a prototype of metabolic architecture. An elementary node of ARM

S. Yasui
Graduate School of Information Sciences and Arts, Toyo University, Kawagoe, Japan
e-mail: s3B101600087@toyo.jp

M. Uehara (✉)
Department of Information Sciences and Arts, Toyo University, Kawagoe, Japan
e-mail: uehara@toyo.jp

© Springer International Publishing AG 2018

215

R. Lee (ed.), *Computational Science/Intelligence and Applied Informatics*,
Studies in Computational Intelligence 726, DOI 10.1007/978-3-319-63618-4_16

Linux Cloud (ALC) is Linux installed in an ARM Single Board Computer (SBC). However, currently, it is necessary to exchange elements manually.

In order to exchange components easily, we employ uniform parts. We employ Raspberry Pi [5] as an ARM SBC. In order to create a Cloud using ARM, it is necessary to virtualize both hardware and network. Although ARM has no hardware-supported virtualization, hardware can be replaced using LXC(Linux Container). Also, we will configure a network of ALC in mesh form. Then we will virtualize the network with OpenFlow [6]. In this research, the network is called OpenFlow mesh network. The client uses the host on the mesh network as a Cloud. In order to set up routing between the client and the host, we need to grasp the topology of the mesh network.

Moreover, metabolic architecture maintains the latest condition by exchanging components. When exchanging parts, internal information of the topology is changed. In order to maintain an existing state, the routing changes at the same time as the topology changes. Therefore, a system needs to implement the function of grasping the topology with OpenFlow and flexibly changing the routing.

Section 2 explains related research and Sect. 3 outlines the implementation technology and applications to be used. Section 4 describes requirements for the system and Sect. 5 explains the implementation. Finally, Sect. 6 concludes and describes future works.

2 Related Works

2.1 Metabolic Architecture

According to [3] and [4], metabolic architecture is metabolic type calculation architecture. It is an architecture in which metabolic architecture is able to exchange components dynamically. It is possible to maintain through exchange of faulty and aging components for minimum cost. Moreover, keeping components up to date improves sustainability and reliability. Metabolic architecture has the following five components, and a conceptualization of the architecture is shown in Fig. 1.

First is the Metaboloid, a small computer which has no inner battery. The Metaboloid has communication terminals extending in 6 directions, and communicates with other Metaboloids using the communication terminal. We implemented the same communication function via an Ether adapter on a USB hub connected to Raspberry Pi.

Second is the Slot. This is the part that stores multiple Metaboloids. The Slot is used to connect and carry the Metaboloid.

Third is the Power Queue (PQ). This is calculation space and supplies power to the Slot. PQ has multiple Slots, and replaces Slots at a regular interval. This replacement is the metabolic aspect of the architecture.

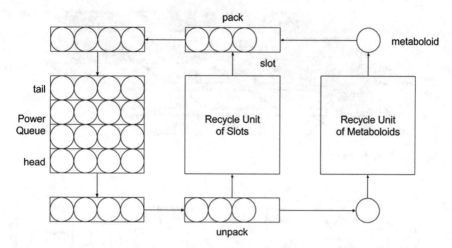

Fig. 1 Metabolic Architecture

Next is the Recycle Unit (RU). This functions with both Slot and Metaboloid. It collects waste material that is the Slot and Metaboloid from the PQ. Moreover, it confirms whether to reuse them. When reuse is possible, they carry them to the Delivery System.

Finally is the Delivery System (DS). The delivery system joins each component in the architecture. It sends Slot, in which Metaboloid is stored, to the PQ. The DS picks up Metaboloid from the Slot within the PQ and sends it to the RU.

In this paper, we use Metaboloid, Slot, and Power Queue in order to realize ALC.

2.2 ARM Linux Cloud

ALC physically consists of two components. The first is a mesh network of a two-dimensional plane. A switch on the plane point (x,y) is shown as SWxy0. We are calling the mesh network the 2D Mesh.

The second network is switches of an array connected on the 2D Mesh. We are calling this the 1D Mesh, and the switch in the 1D Mesh is shown as SWxyz(z > 0). Each switch in the same 1D Mesh will directly connect to other switches using their own weight. 1D Mesh is independent in the XY-axis. Therefore, the switch directory communicates only with others in the same 1D Mesh. For example, when SW111 communicates with SW121, a signal needs to go through between SW110 and SW120 on the 2D Mesh.

The connector between 1D Mesh and 2D Mesh is limited to one place. Where replacing the 1D Mesh, the connection of both components is released. After that, the 1D Mesh replaces an alternative one. Consequently, the architecture allows easy replacement of the 1D Mesh. Moreover, the switch in the 1D Mesh is set up as a

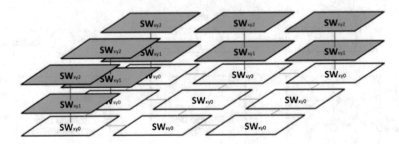

Fig. 2 ALC Architecture

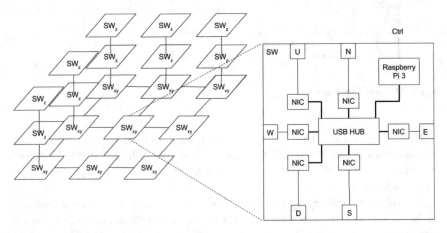

Fig. 3 The structure of openFlow mesh switch

Linux Container as a virtualization application for hardware. We supply the Linux Container as a virtual server.

In this structure, the component of the ALC is corresponding to that of the metabolic architecture. Raspberry Pi is the Metaboloid, the 1D Mesh is the Slot, and the 2D Mesh is the PQ. The ALC architecture is shown in Fig. 2.

Switches which build the 1D Mesh and the 2D Mesh are Open vSwitch software implemented with Raspberry Pi. All switches have the same structure. The USB hub connected to Raspberry pi has 6 USB NICs. Therefore, the switch may use six ports. Figure 3 below shows the details of the switch.

SWxy0 that makes up the 2D Mesh uses 5 ports. One of these is the Up port (U), which connects to the 1D Mesh. The other ports are for connecting to other SWxy0 destinations, which are North port (N), South port (S), East port (E) and West port (W). N connects to SWx(y+1)0, S connects to SWx(y-1)0, E connects to SW(x-1)y0, W connects to SW(x+1)y0, and U connects to SWxy1.

Each switch has an identification number (id). Therefore we take up 5 switches connected to SWxy0 and gather the connection information of the switches on the 2D Mesh. SWxyz that makes up the 1D Mesh uses 2 ports. First is the Up port (U), which connects to the SWxy(z+1) in the same 1D Mesh. Second is the Down port (D), which connects to the SWxy(z-1) or the switch in the 2D Mesh. In the 1D Mesh, we take up 2 switches connected to the U and D ports. Where we take up the connection information of SWxy1, we need to get the id of SWxy0 connecting to D, and of SWxy2 connecting to U.

In order to connect to the Open Flow controller, the switch uses the wireless connection on Raspberry Pi 3.

2.3   SDN (Software Defined Network)

SDN is a concept aimed at controlling networks that constitute and establish the settings for software. There is an OpenFlow protocol as a technology for realizing SDN. It divides existing network functions into planes for each role. First is a controller which is the control plane for routing, second is a virtual switch which is the forwarding plane for data. They communicate various information using a protocol called OpenFlow message. The software we use in this study is listed below.

Open vSwitch [7] is the virtual switch software based on OpenFlow switch specification [8] and generates switch server instances. Each switch has an original identification number and has a flow table and a flow entry (hereafter, entry) used for packet forwarding. The table stores multiple entries. In this paper, we refer to the original identification number as id.

Ryu Framework [9] is the controller framework corresponding to OpenFlow protocol. A process of the switch software such as Open vSwitch will be able to set up with a program flexibly.

2.4   LLDP (Linked Layer Discovery Protocol)

LLDP(IEEE 802.1AB)[10] is a link layer protocol for gathering neighbors. An application grasps a network topology in the mesh network using neighbor discovery with a LLDP packet. The Ryu controller in the application sends the LLDP packet to the switchs ports. The packet contains the id of the switch that will receive it, a port number in the switch to send it to another switch, and TTL (Time to Live).

Moreover the Ryu controller will be able to set up a special configuration for a packet. The controller sets a forwarding function to the LLDP packet. This function is automatically forwarded to the neighbor port when the packet reaches the port.

In order to discover the topology using the LLDP packet, the controller needs to set up a flow entry which the switch that received the LLDP packet has for sending an OpenFlow message to the controller. The process of neighbor discovery using the LLDP packet is shown in Fig. 4 below. However, the flow entry has already been set up.

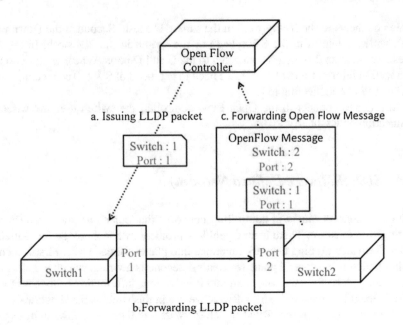

Fig. 4 The process flow of neighbor discovery

(a) The controller issues the LLDP packet to a port of the target switch.
(b) The switch that received the packet forwards it from the switchs port configured in its setting as sender. However, the receiver is, as yet, unknown.
(c) Where the receiving switch is connected to the controller, corresponding flow entry is set up for the switch. When the receiver gets the packet, it sends the OpenFlow message containing the packet to the controller. In the message is the receivers id and port.

The controller gets the receivers id and port number in the message. Moreover, it gets the senders id and port number in the packet encapsulated in the message. Consequently, the controller grasps a link between switches.

3 OpenFlow Mesh

3.1 Components

OpenFlow Mesh [11] is configured by the mesh network consisting of the switch corresponding to OpenFlow and the controller to manage the switch process. The switch on the mesh network connects multiple joined switches and the controller.

The switch divides data forwarding and routing control functions using Open-Flow. The Open vSwitch as the switch on the mesh network forwards a packet. Ryu controller manages the routing control.

3.2   Routing

When setting up a routing between SWxy1 and SWxy2 on the mesh network, two switches exist in the same 1D Mesh. Therefore a virtual network is configured for routing SWxy1–SWxy2. However, when setting up a routing between SWxy1 and SW(x+1)y1, two switches exist in a different 1D Mesh. So, a virtual network needs to be configured using switches on the 2D Mesh. Where showing the least routing between SWxy1 and SW(x+1)y1, the virtual network is configured on SWxy1 - SWxy0 - SW(x+1)y0 - SW(x+1)y1. SWxy0-SW(x+1)y0 on the 2D Mesh, and multiple routing exists. If the least routing on SWxy0 - SW(x+1)y0 is disconnected, the mesh network will be able to configure another routing. This aspect can be said of all routing on the 2D Mesh. Owing to this, we anticipate a certain level of fault tolerance in routing on the mesh network.

4   Implementations

In order to realize a metabolic system in ALC, the mesh network needs to have the following functions listed below.

First is a function which grasps the network topology information and keeps it. ALC uses the topology of all networks for setting up routing. However, ALCs network always changes its topology. Owing to this, an application takes up topology at a regular interval, and keeps the latest topology information.

Second is the fault awareness function. When the network topology is changed, the topology information needs to be renewed in the application. Where a part of the network is not used, the application specifies the faulty part. After that, the application renews information of the faulty part on the topology. Therefore, the application prevents routing using mistaken topology information.

Below we present classes of an application implemented corresponding to a role. The Ryu controller is the Controller, an application to maintain the network topology is the Topology, and an application that distributes information obtained by the Controller for each application is called the Central.

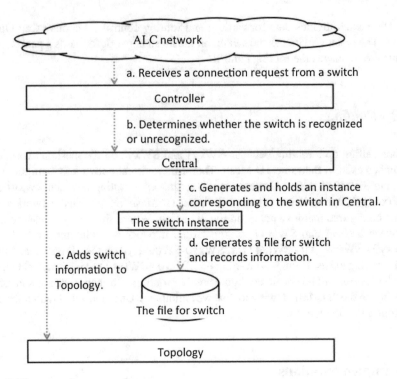

Fig. 5 The connection process flow

4.1 Self-Recognition

By always grasping latest topology in ALC, the application flexibly sets up routing for packet forwarding. In order apprehend changes in the network topology, the application needs to implement two functions: first, to take up connection information between the switch and the controller; and second, to take up link information between switches on the network.

4.1.1 The Connection of the Switch and the Controller

First is the function of taking up connection information between the switch and the controller. When the switch establishes the connection to the controller, switch information is generated and held in the application. Moreover, switch information is registered to Topology, which holds network topology information. The connection process of the flow is shown in Fig. 5.

(a) The Ryu controller receives a connection request from a switch.
(b) The application confirms that information for this switch is held in the Central. Where the Central does not hold switch information, the application determines a newly connected switch for the controller.
(c) When the Central does not hold switch information, the application generates a switch instance in the Central. The instance contains information of its id, its ports, link and applied flow entries. In addition, each switch instance sets a confirmation standby time for survival and response at generation. At the same time, the application adds a flow entry to the corresponding switch on the ALC network. This flow entry ensures that the switch returns response to the controller, when the Controller sends the packet for survival confirmation to the switch.
(d) The application generates a file for the corresponding switch in a special directory in the hardware.
(e) The application adds target switch information as a node to the Topology.

4.1.2 Grasping the Network Topology

Second is the function of dynamically taking up a link between switches. In order to grasp the topology information of the ALC network at regular intervals, the controller performs a neighbor discovery using the LLDP packet to the switch in the network. The process flow of taking up the link using the LLDP packet is shown in Fig. 6.

(a) At regular intervals, each switch instance on the Central requests that the Controller issue an LLDP packet.
(b) The Controller generates the packet and sends it to the corresponding switchs ports on the network.
    When the LLDP packet reaches the switchs ports, a link to the port is taken up in the ALC network. The flow of the neighbor discovery is the same as in Fig. 4.
(c) The Controller receives an OpenFlow message that contains the LLDP packet from a switch which is received from the network containing the LLDP packet.
(d) The Controller parses the received OpenFlow message and gets out link data in it. In addition, the Controller sends link data to the Central. The Controller specifies a switch changing information from link data.
(e) The Central applies link information to each switch instance.
(f) Where a certain switch instance has changed its link information, the instance renews a file corresponding to itself.
(g) Where the switch instance changes link information, the Central makes the Topology load the changed information in the file.
(h) The Topology loads the file and changes the internal information.

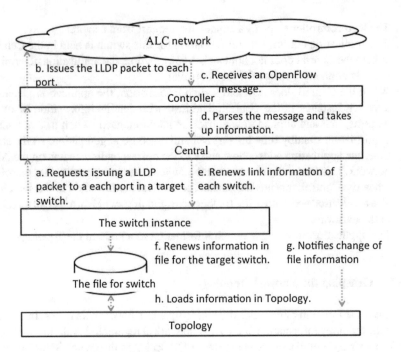

Fig. 6 The process of taking up the link

4.2 Fault Awareness

The application detects when a fault has occurred in the ALC network, and obtains a topology change. Faults might include a link disconnecting, or a switch shutting down. When the application detects the fault, it re-grasps the topology and renews information. In order to renew information, the application implements two functions. First is a processing for an OpenFlow event corresponding to the topology change. Second is the method that confirms a switch survival from the application.

4.2.1 Grasping for The Topology Change Event

An event is a link status change between switches. The Controller parses the notified event, and renews information using change data in the application. Unlike taking up the inter-switch link, the function passively detects the network topology change. The process for the topology change event is shown in Fig. 7.

(a) The switch connected to the Ryu controller transmits topology information such as a changing port and link status.

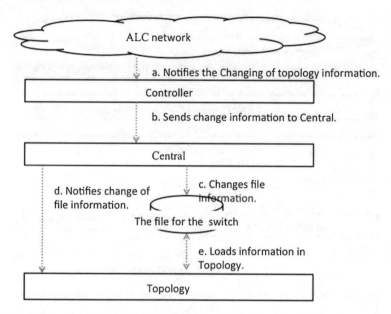

Fig. 7 The process for the topology change event

(b) The Controller sends information to the Central. If it is information for topology changes, the Central calls for the method for changing information in a target switch instance.
(c) The target switch instance changes information.
(d) Where the link information changes, the target switch instance renews the file information corresponding to it.
(e) Simultaneous with the output, the Topology reflects information to the file of changing switch. Moreover the Central notifies the Topology in order to load the file.
(f) The Topology that received the notification loads the target file, and changes the information.

4.2.2 Monitoring the Switch Survival

This is the function in which the Ryu controller confirms a switch survival. The topology change detection uses notification which the controller sends to the switch. However, when the switch is stopped, this function cannot be used. As mentioned above in Sect. 5 each switch instance holds the confirmation standby time for survival. After the confirmation standby time for survival, each switch instance executes a survival monitoring for the switch. Figure 8 is a flow of the process for the survival confirmation.

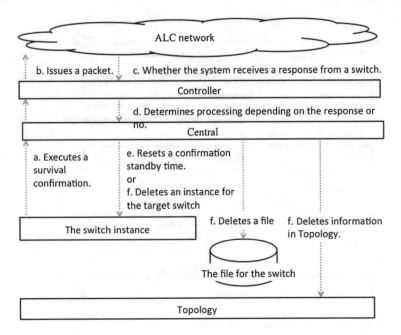

Fig. 8 The process for the survival confirmation

(a) After confirmation standby time has finished, each switch instance executes the survival confirmation for the switch in the ALC network and sets up a response standby time at the same time.

(b) The application issues a packet for the response from the target switch.

(c) Where the Controller receives the response, it sends notification to the Central.

(d) If the response returns within the standby time, the Central executes the process e. However, if the response does not return within the standby time, the Central executes the process f.

(e) Where the response exists, each switch instance resets the confirmation standby time. This process is repeated.

(f) Where the response does not exist, the application deletes the instance, file and information in the Topology of the target switch.

5 Evaluations

Here, the implemented functions are evaluated. Figure 9 shows a topology for evaluations. In addition, an executing result of each function is shown.

The id in Fig. 9 is the identification number of a switch. The application uses the id to identify the switch. A network link is shown as a blue dotted line. A connection between the switch and a controller is shown as a red dashed line.

Fig. 9 The openflow mesh
in evaluations

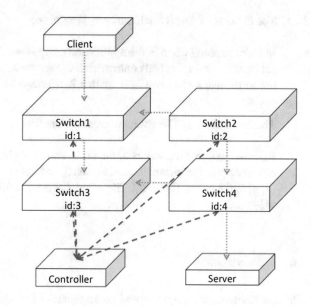

5.1 Self-Recognition

Here, we evaluate the Evaluation of the Network Grasping function. When the switch
connects the controller, we confirmed that a switch instance is generated and registered to the Topology in an application.

1. The Process of The Switch Connection
   We confirmed that the application grasps the id of 1, 2, 3 and 4 switches connected
   to the controller.
2. The Grasping The Network Link
   The links (1, 2), (1, 3), (2, 4) and (3, 4) exist in the network. We confirmed that
   the Topology grasps these links using the LLDP packet.

5.2 Fault Awareness

we evaluated the fault awareness function. A switch 2 battery stops in the topology
of Fig. 9 as a fault. We confirmed that the application detected the topology change
when the switch stopped, as well as changing the information.

1. The Process of The Switch Survival Monitoring

   In order to stop switch 2, the application grasps switches 1, 3, and 4. The application executes a survival confirmation to each switch at a set time, and in order not to receive a response from switch 2, it grasps switches with the id of 1, 3, and 4.

2. The Evaluation of The Fault Awareness Function

   Because switch 2 has stopped, the (1, 2) and (2, 4) links are deleted. The application deletes the information of switch 2, and also information of the (1, 2) and (2, 4) links in the Topology. Therefore, the application preserves the (1, 3) and (3, 4) links.

6 Conclusions

In this research, we implemented an application that detects changes of network topology and grasps a new network automatically. Thus, it will be possible to recover automatically from network failures. As a result, sustainability of the ALC network could realize.

In the future, we will implement a function which will grasp a host connected to the network. We will also install LXC as a virtual host. In current methods of grasping topology using LLDP, the entry depends on the OpenFlow framework. Thus, it is impossible to grasp hosts and clients that do not support OpenFlow. Therefore, the routing path may not contain the client and the host. We need an alternative way of observing the link between a switch and a client directly connected to the client and another link between a switch and a host directly connected to the client and another link, For these reasons, we need a system that grasps switches to which the host and the client are directly connected.

References

1. Uehara, M.: Research trends on renewable computing system. In: Reports from WReCS-2013 Workshop-IEICE TR RIS No. 8, vol. 6, pp. 1–6 (2013) (2013.10.19) (in Japanese)
2. Nakada, Takashi, Nakamura, Hiroshi (Eds.): Normally-Off Computing, Springer (2017)
3. Uehara, M.: Metabolic Computing. In: Proceedings of the 5th International Workshop on Advanced Distributed and Parallel Network Applications(ADPNA2011) in conjunction with the 14th International Conference on Network-Based Information Systems(NBiS2011), pp. 370-375 (2011) (2011.9.7-9,Tirana,Albania)
4. Uehara, M.: "Metabolic Computing: Toward Truly Renewable Systems". IGI Global, Int. J. Distrib. Syst. Technol. (IJDST), 3(3), pp. 27-39 (2012) (July-September 2012)
5. Raspberry Pi Foundation: Raspberry Pi - Teach, Learn, and Make with Raspberry Pi. https://www.raspberrypi.org/

6. McKeown, Nick, et al. "OpenFlow: enabling innovation in campus networks." ACM SIG-COMM Comput. Commun. Rev. **38**(2), 69-74 (2008)
7. Open vSwitch: http://openvswitch.org/
8. Open Networking Foundation: Open Flow Switch Specification. https://www.opennetworking.org/images/stories/downloads/sdn-resources/onf-specifications/openflow/openflow-switch-v1.5.0.noipr.pdf
9. Ryu SDN Framework Community: Ryu SDN Framework. https://osrg.github.io/ryu/
10. Congdon, P.: Link Layer Discovery Protocol and MIB ver.0.0. http://www.ieee802.org/1/files/public/docs2002/lldp-protocol-00.pdf
11. Uehara, M., et al. : "OpenFlow Mesh for Metabolic Computing". In: Barolli, L. (eds.) Advances on Broad-Band Wireless Computing, Communication and Applications, pp. 613-620 In: Proceedings of the 18-th International Symposium on Multimedia Network Systems and Applications (MNSA-2016) in conjunction with the 11-th International Conference On Broad-Band Wireless Computing, Communication and Applications (BWCCA-2016), (November 5–7, 2016, Soonchunhyang University, Asan, Korea)

Author Index

© Springer International Publishing AG 2018
R. Lee (ed.), *Computational Science/Intelligence and Applied Informatics*,
Studies in Computational Intelligence 726, DOI 10.1007/978-3-319-63618-4

Printed in the United States
By Bookmasters